INVASIVE ALIEN SPECIES:
A Toolkit of Best Prevention and Management Practices

Edited by

Rüdiger Wittenberg and **Matthew J.W. Cock**

Global Invasive Species Programme(GISP)

 SCOPE   CAB *International*

CABI Publishing on behalf of the Global Invasive Species Programme

Global Invasive Species Programme(GISP)

SCOPE   CAB *International*   **IUCN**
The World Conservation Union

Published by: CAB International, Wallingford, Oxon, UK,
The views expressed in this publication do not necessarily reflect those of CAB International.
First printed 2001. Reprinted 2003
Printed and bound by the Cromwell Press, Trowbridge, Wiltshire

Citation: Wittenberg, R., Cock, M.J.W. (eds.) 2001. Invasive Alien Species: A Toolkit of Best Prevention and Management Practices. CAB International, Wallingford, Oxon, UK, xii - 228.

ISBN: 0 85199 569 1
Catalogue records of this book are available from the British Library, London, UK, and from the Library of Congress, Washington DC, USA

Design: The Visual Group, 345 California Avenue, Palo Alto, CA 94306, USA.
Tel: +01 650 327 1553, Fax: +01 650 327 2417 Email: visual@batnet.com

Available from: CABI Publishing, CAB International, Wallingford, Oxon OX10 8DE, UK
Tel: +44 (0)1491 832111, Fax +44 (0)1491 833508, Email: cabi@cabi.org
Website http://www.cabi.org
CABI Publishing, 10 East 40th Street, Suite 3203, New York, NY, 10016, USA
Tel: +1 212 481 7018, Fax +1 212 686 7993, Email: cabi-nao@cabi.org

Cover images: *Front cover*: Top left: Water hyacinth, *Eichornia crassipes*, an infestation in Antananarivo, Madagascar (R.H. Reeder, CABI Bioscience). Right: Staff of Malawi Fisheries Department programme rearing water hyacinth biological control agents, Malawi Fisheries Department with support from CAB International and the Plant Protection Research Institute, Republic of South Africa (M.J.W. Cock, CABI). Background: *Neochetina eichhorniae*, a biological control agent for water hyacinth (CABI Bioscience). Lower left: Water hyacinth infesting a lagoon village in Benin (D. Moore, CABI Bioscience). Lower foreground: Flower of water hyacinth (R.H. Reeder, CABI Bioscience). *Back cover*: Clearing of invasive alien trees in the Western Cape Province (Working for Water Programme, Republic of South Africa); Argentine ant, *Iridomyrmex humilis* (SA Museum, H. Robertson); Rat, *Rattus* sp. (Jack Jeffrey Photography); American comb jelly, *Mnemiopsis leidyi* (Harbison); Brown tree snake, *Boiga irregularis* (Gordon H. Rodda/USGS).

The Global Invasive Species Programme (GISP) is co-ordinated by the Scientific Committee on Problems of the Environment (SCOPE), in collaboration with the World Conservation Union (IUCN), and CAB International (CABI). GISP has received financial support from the United Nations Environment Programme (UNEP) - Global Environment Facility (GEF), United Nations Education, Scientific and Cultural Organization (UNESCO), the Norwegian Government, the National Aeronautics and Space Administration (NASA), the International Council for Scientific Unions (ICSU), La Foundation TOTAL, OESI, the David and Lucile Packard Foundation, and the John D. and Catharine T. MacArthur Foundation. Participating groups and individuals have made substantial in-kind contributions. GISP is a component of DIVERSITAS, an international programme on biodiversity science.

CONTENTS

PREFACE

The Global Invasive Species Programme (GISP) is co-ordinated by the Scientific Committee on Problems of the Environment (SCOPE), in collaboration with the World Conservation Union (IUCN), and CAB International (CABI). GISP has received initial financial support from the United Nations Environment Programme (UNEP) - Global Environment Facility (GEF), United Nations Education, Scientific and Cultural Organization (UNESCO), the Norwegian Government, the National Aeronautics and Space Administration (NASA), the International Council for Scientific Unions (ICSU), La Fondation TOTAL, the David and Lucile Packard Foundation, and the John D. and Catharine T. MacArthur Foundation. Participating groups and individuals have made substantial in-kind contributions. GISP is a component of DIVERSITAS, an international programme on biodiversity science.

The overall aim of GISP is to assemble the best available data on various components of the invasive alien species problem. This manual is one of the tools produced by GISP Phase I efforts.

The toolkit was designed and partially drafted at an international workshop held in Kuala Lumpur, 22-27 March 1999, in conjunction with the GISP Early Warning component. The participants of the workshop are listed in the opening pages. Working from this excellent beginning, Rüdiger Wittenberg and Matthew Cock of CAB International prepared the text of the toolkit, which was then reviewed by the participants of the Kuala Lumpur workshop and their feedback incorporated. Dick Veitch of New Zealand acted as a third editor during this review process. The resultant draft was then provided to participants at the GISP Phase I Synthesis Conference held in Cape Town, Republic of South Africa in September 2000. Further review and input by the participants at the Conference was received. Many of the valuable suggestions made at GISP Phase I Synthesis have been incorporated, and the "final" toolkit text prepared for publication. The text and case studies will be adapted to form a website, which is intended as a dynamic version of the toolkit, to be updated with new information, internet links, and case studies as they become available.

The Kuala Lumpur workshop discussed to whom the toolkit should be directed, and concluded that the main focus should be to assist those involved in environment and biodiversity conservation and management. It is not aimed directly at the public, policy makers, quarantine services etc., but should provide insights for these groups in addition to conservation managers. Nevertheless, the contents are likely to provide useful information to a wider group and will be widely disseminated.

The workshop also discussed whether the toolkit should try and address all types of invasive species (e.g. agricultural, forestry, human health, etc.) or just those that affected environment and biodiversity. It was concluded that:

➤ Human diseases, although technically invasive, fall outside the scope of the toolkit, and are well addressed by other means;

➤ Examples, case histories and lessons of best practice will inevitably come from traditional sectors such as agriculture, forestry, etc.;

➤ Many of these invasive species will also have a significant impact on and be interlinked with the environment and biodiversity;

➤ The case for motivating funding sources will often depend more on the economic impact of invasive species in terms of increased production costs, lost production, loss of ecosystem services, human health etc.

The toolkit is intended to be global in its applicability, although there is a small island focus, recognizing that the impact of invasive alien species on biodiversity is greater in small island systems. In any case, we anticipate that to be most useful and effective, the toolkit will need to be locally adapted for different countries or regions (Chapter 6). In this regard we would like to note that the case studies represent the particular expertise of the workshop participants, and the people we were subsequently able to work with during the preparation of the toolkit, and are therefore not representative of the full range of experience worldwide. We recommend additions of nationally and regionally focused case studies in local adaptations of the toolkit.

In designing the toolkit, the Kuala Lumpur workshop also considered whether the GISP toolkit should be restricted to invasive alien species or also try to cover invasive indigenous species. It was concluded that:

➤ There are several examples of important indigenous invasive species, usually linked (or suspected to be linked) to land use change;

➤ Large parts of the toolkit would be irrelevant to this type of problem (e.g. most of early warning and prevention), although significant parts would be potentially useful (e.g. much of management);

➤ On balance the toolkit should retain its focus on invasive alien species, but where relevant the text should address what was or was not relevant to invasive indigenous species.

The layout of the toolkit is intended to be largely self-explanatory. An introductory chapter to set the scene is followed by Chapter 2 on building strategy and policy, (i.e. how to develop national plans and support for them). Methods for prevention of invasive species and the risk-analysis process are dealt with in Chapter 3, while methods for early detection of new invasive species are reviewed in Chapter 4. A broad review of different management approaches is offered in Chapter 5, and some thoughts on how to use the toolkit are provided in Chapter 6. In the text we have recognized that there are often fundamental differences in the approach depending on the ecosystem being invaded (terrestrial, freshwater, marine) and the taxonomic group of the invasive species (vertebrates, invertebrates, diseases, plants, etc.). We have attempted to keep these distinctions clear by use of section headings.

During the course of preparation of the toolkit, we soon recognized that it is difficult to draw useful generalizations and make predictions due to the complexity of this global problem. These complexities involve the various traits of completely different taxonomic groups, in different localities, and different ecosystems, that are affected by different human activities. Thus, we chose to illustrate the text with case studies of successful projects and examples highlighting key problems. The immense scope of the field of prevention and management of invasive alien species makes it impossible to include all aspects in depth in one manageable toolkit. Hence, the actual scope of the toolkit is not so much a "how to" document, but a "what to do" document, with case studies to provide insights into how one might approach an invasive alien species issue. This document provides advice on what to do and where to look for more information.

We need to make two editorial provisos for users of the toolkit. Whenever you find the word "species", it is not necessarily meant in the strict scientific sense but may go beyond this to include other taxonomic levels. An alien subspecies can be equally as alien and different in an area as an alien species. Moreover, the status of super-species, species and subspecies is often debatable. Similarly the term "national" needs to be broadly interpreted (in light of local circumstances) as meaning regional, national, or sub-national where appropriate. In some instances, it could be as usefully applied to an ecological area as a political unit.

We gratefully thank all of the individuals who have provided inputs to this toolkit, including the GISP team around the world. In particular, the participants at the Kuala Lumpur workshop, international specialists who provided case studies, information, and reviewed the text, and, finally to all who provided inputs during and after the GISP Phase I Synthesis Conference. Following input of the Synthesis conference, summary sections based on other components of GISP have been added to cover the GISP database, human dimensions of invasive alien species, marketing strategy and legal frameworks. We particularly thank Alan Holt (The Nature Conservancy) and Nattley Williams (IUCN Environmental Law Centre) for the inputs they provided on the last two topics respectively.

If this toolkit is useful and valuable, much of the credit should go to all those who provided the design, information and content that went into it. However, the responsibility for producing the final text has been that of Rüdiger Wittenberg and Matthew Cock, and if we have incorporated errors, failed to understand the information provided, or not risen to the challenges of some the excellent suggestions provided, then the fault lies with us. Nevertheless, we believe this toolkit provides a view of the state of the art at the beginning of the new millennium. We think it will be useful to many individuals and countries and we recommend it to you.

Matthew Cock and Rüdiger Wittenberg

The Kuala Lumpur Workshop participants who designed this toolkit:

Dr. Ahmed Anwar Ismail
Centre for Strategic Research, Environment
and Natural Resource Management
Malaysian Agricultural Research and
Development Institute (MARDI)
P.O. Box 12301,
50774 Kuala Lumpur
Malaysia

Ms. Yvonne C. Baskin
Science Writer
200 So. 23rd Ave.
Bldg. D7, #145
Bozeman, Montana 59718
USA

Dr. Mick N. Clout
Chair, IUCN Invasive Species Specialist
Group
School of Biological Sciences/SEMS
University of Auckland
PB 92019, Auckland
New Zealand

Dr. Matthew J. W. Cock
Centre Director
CABI Bioscience Centre, Switzerland
1 Rue des Grillons
CH-2800 Delémont
Switzerland

Dr. Lucius G. Eldredge
Executive Secretary, Pacific Science
Association
Bernice P. Bishop Museum
Honolulu, Hawaii 96817
USA

Dr. Simon V. Fowler
Insect Ecologist
Landcare Research - Manaaki Whenua
Private Bag 92170
Mt Albert, Auckland
New Zealand

Dr. John W. Kiringe
Team Leader
Biology of Conservation Group
Department of Zoology
University of Nairobi
P.O. Box 30197 Nairobi
Kenya

Dr. Lim Guan Soon
CABI SE Asia Regional Centre
P.O. Box 210
43 409 UPM Serdang, Selangor
Malaysia

Dr. Loke Wai Hong
CABI SE Asia Regional Centre
P.O. Box 210
43 409 UPM Serdang, Selangor
Malaysia

Ms. Sarah Lowe
School of Environmental and Marine
Sciences
University of Auckland
Private Bag 92019, Auckland
New Zealand

Dr. R. K. Mahajan
Senior Scientist
National Bureau of Plant Genetic Resources
Pusa Campus, New Delhi – 110 012
India

Dr. John R. Mauremootoo
Plant Conservation Manager
Mauritian Wildlife Foundation
Black River Office
Avenue Bois de Billes
La Preneuse
Mauritius

Dr. Jean-Yves Meyer
Delegation a la Recherche
B.P. 20981 Papeete
Tahiti, French Polynesia

Mr. Yousoof Mungroo
Director, National Parks and Conservation Service
Ministry of Agriculture, Food Technology & Natural Resources
Reduit
Mauritius

Mr. Anisur Rahman
UNEP- Biodiversity Programme
Box 30552
Nairobi, Kenya

Dr. John M. Randall
The Nature Conservancy (TNC)
Invasive Species Program, Dept. of Vegetable Crops & Weed Science
124 Robbins Hall
University of California
Davis, CA 95616
USA

Mr. Selby Remie
Senior Conservation Officer
Division of Environment
Ministry of Environment and Transport
Botanical Gardens, Mont Fleuri
P.O. Box 445, Victoria, Mahé
Republic of Seychelles

Dr. Soetikno S. Sastroutomo
CABI SE Asia Regional Centre
P.O. Box 210
43 409 UPM Serdang, Selangor
Malaysia

Dr. Greg H. Sherley
Programme Officer, Avifauna Conservation and Invasive Species
South Pacific Regional Environment Programme
PO Box 240, Apia
Samoa

Prof. Daniel S. Simberloff
Dept. Ecology & Evolutionary Biology
University of Tennessee
480 Dabney / Buehler
Knoxville, Tennessee 37996
USA

Mr. Jim Space
Pacific Island Ecosystem at Risk (PIER)
USDA Forest Service
Institute of Pacific Islands Forestry
Honolulu, Hawaii
Mailing address:
11007 E. Regal Dr.
Sun Lakes, AZ 85248-7919USA

Mr Philip A. Thomas
Research Associate / Computer Specialist
Hawaiian Ecosystems at Risk Project (HEAR)
P.O. Box 1272
Puunene, Hawaii 96784
USA

Dr. Brian W. van Wilgen
CSIR Division of Water, Environment and Forestry Technology
P.O. Box 320
Stellenbosch 7599
South Africa

Mr. Dick Veitch
48 Manse Road
Papakura
New Zealand

Prof. Jeff K. Waage
Chief Executive
CABI Bioscience
Silwood Park, Buckhurst Road
Ascot, Berks SL5 7TA
UK

Mr. Rüdiger Wittenberg
CABI Bioscience
c/o CSIRO European Laboratory
Campus International de Baillarguet
F - 34980 Montferrier sur Lez
France

TOOLKIT SUMMARY

Invasive alien species are recognised as one of the leading threats to biodiversity and also impose enormous costs on agriculture, forestry, fisheries, and other human enterprises, as well as on human health. Rapidly accelerating human trade, tourism, transport, and travel over the past century have dramatically enhanced the spread of invasive species, allowing them to surmount natural geographic barriers. Not all non-indigenous species are harmful. In fact the majority of species used in agriculture, forestry and fisheries are alien species. Thus, the initial step in a national programme must be to distinguish the harmful from the harmless alien species and identify the impacts of the former on native biodiversity.

Development of a national strategy summarizing goals and objectives should be the first step in formulating an alien species plan. The ultimate goal of the strategy should be preservation or restoration of healthy ecosystems. An initial assessment, including a survey of native and alien species (and their impacts) will help define the starting-point and serve as a base for comparison as the programme progresses. The support of all stakeholders must be engaged during the entire programme, ideally using a social marketing campaign. Legal and institutional frameworks will define the basic opportunities for prevention and management of invasive alien species. There are four major options (or better, steps) for dealing with alien species: 1) prevention, 2) early detection, 3) eradication, and 4) control (Figure 1).

Prevention of introductions is the first and most cost-effective option. This lesson has been learned the hard way from several cases of highly destructive and costly invasive organisms such as the zebra mussel in the Great Lakes. Had such species been intercepted at the outset, an enormous loss of native species and/or money could have been prevented. Exclusion methods based on pathways rather than on individual species provide the most efficient way to concentrate efforts at sites where pests are most likely to enter national boundaries and to intercept several potential invaders linked to a single pathway. Three major possibilities to prevent further invasions exist: 1) interception based on regulations enforced with inspections and fees, 2) treatment of material suspected to be contaminated with non-indigenous species, and 3) prohibition of particular commodities in accordance with international regulations. Deliberate introductions of non-indigenous species should all be subject to an import risk assessment.

Early detection of a potential invasive species is often crucial in determining whether eradication of the species is feasible. The possibility of early eradication or at least of effectively containing a new coloniser makes investment in early detection worthwhile. Early detection in the form of surveys may focus on a species of concern or on a specific site. Species-specific surveys are designed, adapted or developed for a specific situation, taking into consideration the ecology of the target species. Site-specific surveys are targeted to detect invaders in the vicinity of high-risk entry points or in high value biodiversity areas.

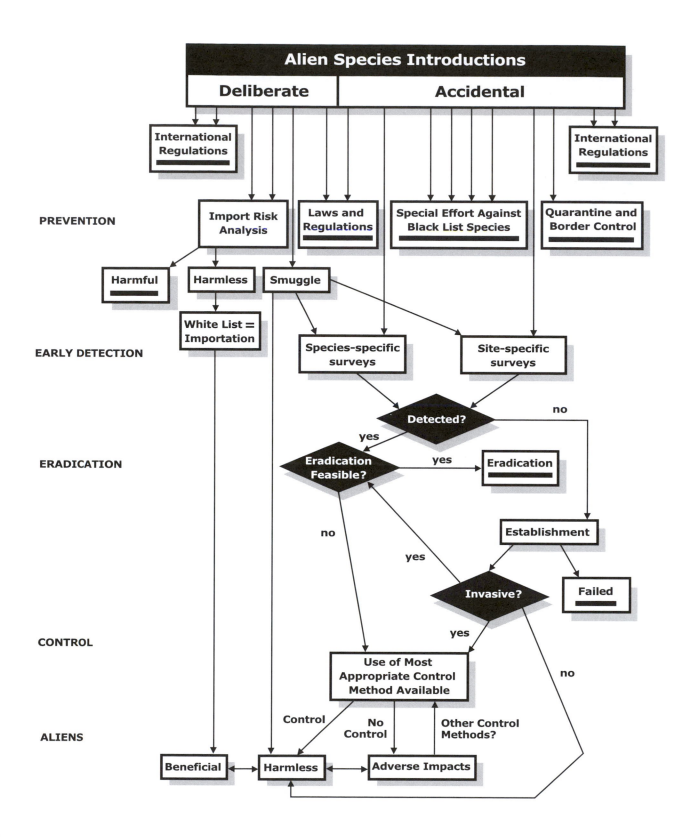

Figure 1 - Summary of options to consider when addressing alien species. Black bars mark the potential final stages of introduced alien species. Diamonds symbolise important bifurcations and decision points.

When prevention has failed, eradication is the preferred course of action. Eradication can be a successful and cost-effective solution in response to an early detection of a non-indigenous species. However, a careful analysis of the costs and likelihood of success must be made, and adequate resources mobilised, before eradication is attempted. Successful eradication programmes in the past have been based on 1) mechanical control, e.g. hand-pulling of weeds or handpicking of snails, 2) chemical control, e.g. using toxic baits against vertebrates, 3) habitat management, (e.g. grazing and prescribed burning), and 4) hunting of invasive vertebrates. However, most eradication programmes need to employ several different methods. Each programme must evaluate its situation to find the best methods in that area under the given circumstances.

The last step in the sequence of management options is the control of an invasive species when eradication is not feasible. The aim of control is to reduce the density and abundance of an invasive organism to keep it below an acceptable threshold. There are numerous specific methods for controlling invasive species. Many of the control methods can be used in eradication programmes, too. Mechanical control is highly specific to the target, but always very labour-intensive. In countries where human labour is costly, the use of physical methods is limited mainly to volunteer groups. Chemical control is often very effective as a short-term solution. The major drawbacks are the high costs, the non-target effects, and the possibility of the pest species evolving resistance. In comparison with other methods, classical biological control, when it is successful, is highly cost-effective, permanent, self-sustaining and ecologically safe because of the high specificity of the agents used. Biological control is particularly appropriate for use in nature reserves and other conservation areas because of its environmentally friendly nature and the increasing instances of prohibition of pesticide use in these areas. Integrated pest management, combining several methods, will often provide the most effective and acceptable control.

Finally, there will be situations where the current techniques for management of invasive alien species are simply inadequate, impractical or uneconomic. In this situation conservation managers may have to accept that they cannot control the invasive alien species and the only recourse is to develop ways to mitigate its impact on key habitats and species. This topic is introduced and discussed briefly, but merits a fuller consideration beyond the scope of this toolkit.

Biological invasions by non-native species constitute one of the leading threats to natural ecosystems and biodiversity, and they also impose an enormous cost on agriculture, forestry, fisheries, and other human enterprises, as well as on human health. The ways in which non-native species affect native species and ecosystems are numerous and usually irreversible. The impacts are sometimes massive but often subtle. Natural barriers such as oceans, mountains, rivers, and deserts that allowed the intricate coevolution of species and the development of unique ecosystems have been breached over the past five centuries, and especially during the twentieth century, by rapidly accelerating human trade and travel (Case Study 1.1 "Acceleration of Colonization Rates in Hawaii"). Planes, ships, and other forms of modern transport have allowed both deliberate and inadvertent movement of species between different parts of the globe, often resulting in unexpected and sometimes disastrous consequences.

Introduced species often consume or prey on native ones, overgrow them, infect or vector diseases to them, compete with them, attack them, or hybridise with them. Invaders can change whole ecosystems by altering hydrology, fire regimes, nutrient cycling, and other ecosystem processes. Often the same species that threaten biodiversity also cause grave damage to various natural resource industries. The zebra mussel (*Dreissena polymorpha*), *Lantana camara*, kudzu (*Pueraria lobata*), Brazilian pepper (*Schinus terebinthifolius*), and rats (*Rattus* spp.) are all economic as well as ecological catastrophes. Invasive non-native species are taxonomically diverse, though certain groups (e.g., mammals, plants, and insects) have produced particularly large numbers of damaging invaders. Thousands of species have been extinguished or are at risk from invasive aliens, especially on islands but also on continents. Many native ecosystems have been irretrievably lost to invasion. Weeds cause agricultural production losses of at least 25% and also degrade catchment areas, near-shore marine systems, and freshwater ecosystems. Chemicals used to manage weeds can further degrade ecosystems. Ballast water carries invasives that clog water pipes, foul propellers, and damage fisheries. Imported pests of livestock and forests reduce yields drastically. Further, environmental destruction, including habitat fragmentation, and global climate change are extending the range of many invaders.

Not all non-native species are harmful. In many areas, the great majority of crop plants are introduced, as are many animals used for food. Some productive forest industries and fisheries are based on introduced species. And introductions for biological control of invasive pests have often resulted in huge savings in pesticide use and crop loss. However, many of the worst introduced pests were deliberately introduced. Horticultural varieties and zoological novelties have become invasive and destructive; fishes introduced for human consumption have extirpated many native species, and even biological control introductions have occasionally gone awry. The rapid development of the science of invasion biology, as well as growing technologies for detecting unintentionally introduced invaders and managing established invasive species, can provide major advances in the war against invasive exotic species, so long as the public and policymakers are aware of them.

A national strategy is required to assess the full scope of the threat of invasive non-native species and deal with it effectively. Also critical to success is a mechanism for international co-operation to stop invasions at their source and to foster the sharing of lessons learned in preventing and dealing with invasions. This toolkit is designed to aid in the elaboration and adoption of an effective national strategy, by pointing to experiences in various nations. The toolkit is written from the perspective of the sustainable use and conservation of biodiversity as embodied in article 8h of the Convention on Biological Diversity, but it addresses a problem that can only be solved through an alliance of environmental, health, industrial, agricultural, and other resource-based sectors of society. Invasive alien species are a development issue. Users will find suggestions on mobilising and generating public support for a national commitment, assessing the current status and impact of invasive exotic species, building institutional support for an effective response to the problem, and putting the strategy on a firm institutional and legal basis.

The toolkit also provides advice, references, and contacts to aid in preventing invasions by harmful species and eradicating or managing those invaders that establish populations. The extensive literature and experience in prevention, eradication, and long-term management can be bewildering and difficult to navigate. The toolkit provides an easy approach into this diverse field.

It is anticipated that the toolkit will not always be directly applicable to all situations. All countries and areas have their constraints, some more severe than others. For example, the constraints facing the small island nations of the Pacific were starkly summarized at the Global Invasive Species Programme: Workshop on Management and Early Warning Systems, held in Kuala Lumpur, in March 1999 (Case Study 1.2 "Particular Problems Related to Invasive Species in the South Pacific"). The toolkit in its present form will require a process of validation to ensure that the content is appropriate and relevant to users. In at least some cases, it will then need to be adapted for local situations, issues and problems.

CASE STUDY 1.1 Acceleration of Colonization Rates in Hawaii

Each year, an average of 20 new alien invertebrates become established in the islands of Hawaii. This is a rate of one successful colonization every 18 days, compared to the estimated natural rate of once every 25-100,000 years. Moreover, in the average year, half of the newly established invertebrates are taxa with known pest potential.

Edited from http://www.hear.org/AlienSpeciesInHawaii/articles/norway.htm "An alliance of biodiversity, agriculture, health, and business interests for improved alien species management in Hawaii" presented at the Norway/UN Conference on Alien Species, July 1-5, 1996, by Alan Holt, The Nature Conservancy of Hawaii, 1116 Smith Street, Suite 201, Honolulu, Hawaii 96817.

CASE STUDY 1.2 Particular Problems Related to Invasive Species in the South Pacific

➤ There are a huge number of islands, many of them tiny and remote, which means wildlife management (especially using sophisticated methods such as those requiring many visits) is practically difficult and expensive.

➤ Because of the small size of many islands and the relatively small number of species that occur there naturally, the potential impact of invasive species on the indigenous biodiversity may be particularly devastating.

➤ The public service infrastructure required to support complex control and eradication operations or border control is not available.

➤ Western standards of communication just do not occur. Telecommunications, let alone access to computers and the internet, are not available on many islands. Shipping and air services to these islands are often infrequent and may be unreliable.

➤ Much of the relevant technical information is in English, which is second language for most Pacific Islanders. Technical English is particularly difficult even for many otherwise bilingual people.

Edited from "Invasive Species in the South Pacific" a paper by Greg Sherley, Programme Officer, South Pacific Regional Environment Programme presented at the Global Invasive Species Programme: Workshop on Management and Early Warning Systems, Kuala Lumpur, Malaysia, 22-27 March 1999

Chapter 2

BUILDING STRATEGY AND POLICY

Summary

This chapter outlines steps involved in building a national invasive alien species strategy. A national strategy needs to be based on an evaluation of the human dimensions of the invasive alien species problem and an assessment of the current situation. All stakeholders must be identified at the beginning of the process and be involved in all phases of preparing the strategy. After an initial assessment, the national strategy must be formulated using all available information, and international co-operation sought where needed for co-ordination of regional efforts and enlisting of external expertise. Relevant legal and institutional elements need to be identified and may need to be further developed to provide a framework for the action plan.

Management strategy and policy must engage the human dimension of the invasive alien species issue. All ecosystems worldwide are disturbed by human activities in one way or another, and people are the main driving force behind introductions of alien species. Since human behaviour has led to most invasions, it follows that solutions will need to influence human behaviour, e.g. by addressing the economic motivations for introductions.

An initial step towards a national strategy should be to identify a cross-sectoral group that will advocate the development of an invasive species initiative. This group will have to gather, assess and present the evidence that invasive species are a major threat to biodiversity in that country and that action needs to be taken. The preparation of an initial assessment is a crucial step. It should include an inventory of existing invasive species, their ecological and economic impacts, and the ecosystems invaded.

The next step should be identifying and involving all stakeholders, and making them aware of the need for a national commitment to address the invasive alien species problem. Key persons in favour of a national strategy need to be strategically involved, and conspicuous invasive species problems in the country can be used to raise public awareness.

An awareness raising campaign should be a central part of a national initiative to educate the public about the problems caused by invasive species and to inform them of the management options available for solving or preventing the problems. "Social marketing" can provide the tools to approach the problem with well-tested techniques to influence human behaviour, and the successful implementation of a social marketing campaign is described in seven steps.

Once the initial assessment is completed and stakeholders have been engaged, the next step is to develop the national strategy. Ideally, a single lead biosecurity agency should be identified or created. If an inter-agency approach is required, responsibilities and tasks need to be clearly defined and allocated between the different agencies.

The vision, goals and objectives for the national strategy need to be established. The ultimate goal is the conservation or restoration of ecosystems to preserve or restore natural biodiversity. The strategy should be integrated with other national initiatives and action plans, and should support a cross-sectoral approach. Based on the information gathered in the initial assessment, priorities for action need to be assigned for prevention and management plans.

Finally, the legal and institutional framework for prevention and management of invasive alien species needs to be considered. Effective management requires appropriate national laws as well as coordinated international action based on jointly agreed standards. Many international agreements address components of the invasive species problem, but national legislation is needed for implementation in each country. On a national level, the first step should be to identify existing relevant legislation and institutions and then identify any gaps, weaknesses and inconsistencies. There are three ways that appropriate national laws could be put in place: a review of existing laws and their consolidation into a single legislative framework, the enactment of one core framework legislation, or the harmonisation of all relevant laws.

Building strategy and policy needs to be based upon a clear understanding of the human dimensions of invasive alien species, their impact and alternatives for their prevention and control. These aspects are repeatedly mentioned throughout this toolkit, but are also the focus of the GISP Human Dimensions group (Case Study 2.17 "Human Dimensions of Invasive Alien Species").

While invasive alien species have important biological implications, the human dimensions will be paramount in achieving solutions. First, the issue has important philosophical dimensions, requiring people to examine fundamental ideas, such as "native" and "natural". Second, virtually all of our planet's ecosystems have a strong and increasing anthropogenic component that is being driven by increasing globalisation of the economy. Even the remotest ecosystems are disturbed by direct or indirect human intervention, which weakens their resistance to invasions. And third, people are designing the kinds of ecosystems they find congenial, incorporating species from all parts of the world.

People are the main driving force in the tremendous increase of organisms moving from one part of the world to another, especially through trade, travel, tourism, and transport. The great increase in the importation of alien species for economic, aesthetic, or even psychological reasons often leads to more species invading native ecosystems, with disastrous results.

Some of the important human dimensions of the invasive alien species problem are historical, philosophical and ethical, economic, cultural and linguistic, health, psychological and sociological, management and legal (cf. Section 2.5.2), military, and the all-important political dimensions. This litany of issues makes it obvious that human dimensions are critically important and that successfully addressing the problem of invasive alien species will require collaboration between different economic sectors and among a wide range of disciplines.

The Convention on Biological Diversity offers member nations an important opportunity for addressing the complex global problems of invasive alien species through improved international co-operation. The human dimensions of invasive species clearly illustrate that it is not species themselves that are driving the problem but the human behaviours which lead to invasions. Thus, a fundamental solution requires looking at the human dimensions of invasions and addressing the ultimate human cause, e.g. the economic motivation that drives species introduction.

2.1   Making the case for national commitment

National strategies and action plans to address the problem of invasive alien species provide an important framework for activities by all parties, governmental and non-governmental. They underline the national commitment to action.

While national strategies are usually formally endorsed by governments, invasive species initiatives in a country often start with persons, groups or institutions who undertake to advocate the development of an invasive species initiative. Thus, in any country an initial step would be to identify a person, group or institution that will advocate the development of an invasive species initiative.

Development of a national strategy could be initiated either by governmental or non-governmental institutions or both. The process can also occur at a regional level, where environmental policy and action is already co-ordinated regionally (Case Study 2.1 "Development of the South Pacific Regional Invasive Species Programme (SPREP)").

2.2   Assessment

What are the initial steps to be taken by the group advocating a national programme? A critical first step is to gather enough information to make the case that confronting the problem of invasive species is a vital element of national biodiversity plans. The preparation of a preliminary assessment based on existing information, which can be accessed from various sources (literature, databases, etc.), will provide an important document on which to base the initiative, as well as a source for comparison later in the process (an exhaustive example of an assessment is provided in Case Study 2.18 "A National Assessment of Invasives: the U.S. Office of Technology Assessment Report"). Crucial information-gathering activities for this preliminary assessment include:

➤ Preparation of an inventory of existing invasive species problems and their known environmental and economic impacts locally, and also impacts reported elsewhere.

➤ Check databases to see if any aliens are in your country or region which are considered serious invasives elsewhere, and consider what kind of impact they might have in your country (see Box 2.1 "Some Internet-Based Databases and Documents on Invasive Alien species").

➤ Besides threats to biodiversity, consideration should be given to threats to ecosystems services, agriculture, forestry, health, and trade. Inclusion of these threats will be particularly important, not only in building a better case but also in identifying stakeholders.

➤ Take into account the various human dimension aspects of invasive alien species (Case Study 2.17 "The Human Dimensions of Invasive Alien Species").

➤ Pay attention to ecosystems that are particularly vulnerable and to endangered species and their habitats.

➤ Identify major pathways for potential future introductions of non-indigenous species, in particular for species known to be pests under similar conditions elsewhere.

➤ Economic analyses are an important and recommended tool as a basic component of an invasives strategy (Case Studies 2.2 "Economic Costs of Invasive Alien Species in the USA", 2.3 "Economic Justification for the "Working for Water" Programme in South Africa", 2.4 "The Economic Case for Control of Leafy Spurge in North Dakota, USA", and 2.18 "A National Assessment of Invasives: the U.S. Office of Technology Assessment Report"). The economics of invasive alien species is the subject matter of another section of GISP, and the GISP website (http://jasper.stanford.edu/gisp) should be monitored for the outputs of this group.

2.3 Building institutional support

Even when an assessment seems to incorporate sufficiently convincing arguments to support a national commitment to address the problem of invasives, the case may often encounter resistance. This may be due to bureaucratic inertia or simply to a lack of capacity to respond. However, it may also fail to motivate decisions-makers who feel it is not their responsibility or priority. It also sometimes happens that these are the same ministries, offices or individuals who were responsible for some introductions of invasive alien species in the past. Such a situation will need careful handling based upon the local culture. In bringing the invasive species problem to the attention of key decision-makers or people of influence, the following considerations should be taken into account:

➤ Identification of the key influential people/organizations, such as scientists, media, politicians, international organizations, etc., who are sympathetic to the invasives species issue and have close relationships with national leaders (Case Study 2.5 "Scientists Petition for Action on Green Seaweed in the USA").

➤ Outside experts often have more success making the same case to a national leader than local experts do (Case Study 2.10 "Co-ordination of Witchweed Eradication in the USA"). Invitations to, and statements by, visiting eminent scientists or media personalities could influence the decision-making process (Case Study 2.6 "Learning from Others' Experience: The case of *Miconia calvescens*").

➤ The initiative could be built on a current crisis in the country to take advantage of public support, e.g. hysteria against zebra mussels (Case Study 3.4 "The Impact of Zebra Mussel on Ecosystems") or Asian longhorned beetles (Case Study 3.5 "Asian Longhorned Beetle, a Threat to North American Forests") in North America, brown tree snake (Case Study 3.14 "Spread of the Brown Tree Snake in the Pacific Region") and *Miconia calvescens* (Case Study 4.6 "Public Awareness and Early Detection of *Miconia calvescens* in French Polynesia") in the Pacific. See also Case Study 2.7 "The Dirty Dozen - America's Least Wanted Alien Species".

Meetings of people and organizations with vested interests in invasive species problems should be held, drawing upon sectors identified in the assessment. These "stakeholders" should identify constraints on national action and prepare plans to address these (Case Study 5.40 "Community-based Aboriginal Weed Management in the 'Top End' of Northern Australia"). The elements of a national strategy will emerge out of a synthesis of these discussions (Case Study 2.8 "Developing a Strategy for Improving Hawaii's Protection Against Harmful Alien Species").

2.4 Social marketing strategies for engaging communities in invasive species management

This section is based upon outputs of the GISP section on education. It presents an overview of how social marketing strategies can be used to promote the issues relating to invasive alien species, and generate the support to address them.

In many places government agencies or non-profit organizations have launched campaigns to raise awareness, but most of these campaigns have not been able to change the behaviour of those whose actions could limit the impact of invasives. An emerging group of campaign planners, who draw from academic research in social sciences and commercial marketing experience, are proposing new techniques that, used in conjunction, can not only raise awareness but also persuade both public and policymakers to act to solve the problem. It can provide the tools to approach the problem systematically, with well-tested techniques in influencing human behaviour. Its goal is to promote behaviours that will improve human health, the environment or other issues with social benefits.

A comprehensive and fully integrated social marketing approach may often not be possible due to a limited budget. During the first phase of any programme it is advisable to consult with a social marketing expert to determine the minimum package required for the achievement of desired results. Since some steps build on others, it is not wise to choose activities randomly. If available resources are inadequate to achieve minimum results, strategic multi-sector partnerships will become important. The first step in the campaign may be to convince potential partners to join and put resources into the campaign.

Social marketing is a step-by-step approach to motivate specific people (often referred to as "stakeholders" or "key audiences") to take some specific, measurable action or actions for the good of the community. It is analogous to commercial marketing, where the objective is to get a targeted set of consumers to buy a specific product.

Public awareness or public education is the work of making people aware of a certain set of facts, ideas, or issues. Social marketing often utilizes public awareness or education campaigns to inform key audiences and predispose them to appropriate action, but takes this process further to get people to act on their new awareness. All too often, campaigns that are intended to cause specific changes in a community stop at simply informing people. This is rarely enough to promote the kinds of specific actions needed to reduce invasive species problems. This is the main characteristic of social marketing.

For more detailed information refer to "A Social Marketing Handbook for Engaging Communities in Invasive Species Management" prepared by The Academy for Educational Development in conjunction with Alan Holt (The Nature Conservancy). In addition there are numerous web sites dedicated to the topic of invasives, which are linked to other invasive web sites, for example www.nbii.gov/invasives.

Social marketing in seven steps:

Step 1: Conduct an Initial Assessment

The success of any social marketing effort depends largely on the quality of its initial assessment. This critical stage will determine future activities. Campaign managers must insure that their own biases do not distort strategies. The assessment step assures that the perspectives of all stakeholders are considered when identifying the key issues.

In trying to answer questions about the characteristics of the invasive, the pathways, and who is involved in the supply of and demand for the invasive (passive or pro-active) with costs and benefits, the problem and the key issues for each stakeholder will be identified. Each stakeholder must be interviewed either individually or collectively to determine what incentives or benefits and potential obstacles to behaviour or policy changes need to be addressed in the social marketing programme.

In addition to identifying all stakeholders, research should begin to identify sources of influence on each group and the various channels of communication through which they might be most effectively reached.

The final assessment document should include:

1. **Situation analysis** - a clear and concise summary of the status of the invasive problem including a statement of the problem, objectives, and strategic options for achieving objectives.

2. **Summaries of interviews** with representatives from each group of stakeholders to understand their particular perspective on or interest in the invasive problem.

3. **An assessment of the potential for partnerships** among stakeholders to address the invasives issue (e.g. areas of specific interest, funding possibilities, complementary resources).

4. **Key issues** - the problems and opportunities that will be encountered in addressing the invasive threat (specific for each group). These issues are identified in the assessment and will be addressed by the marketing strategies. It is important to clearly define what can be done on the supply, demand and policy levels to control the problem; and, which stakeholders can potentially have an impact by taking certain actions. These stakeholders will be your target group. All other groups that might influence the behaviour of these stakeholders become channels through which you can reach your target group.

5. **Potential channels** of communication and influence on stakeholders (interpersonal, electronic, mass media, public relations).

6. **A comprehensive list of recommendations** and potential strategies drawing on outside technical assistance, if necessary.

Step 2: Build a Partnership Task Force

The success of an invasive species social marketing program will depend on the degree to which all key stakeholders are willing to join forces based on common and mutually beneficial objectives. Each participant in the task force will have his/her own motivations and must be educated to understand and appreciate the objectives, motivations, apprehensions and resources of the others. Each of the stakeholders will have other priorities that will draw their attention from the partnership and the campaign. A catalyst must provide continuity and objective expertise in pushing things forward and in providing technical leadership. The task force should be convened by an authoritative and well-respected body, perhaps a government agency or high level official, to ensure responsiveness.

During the first task force meeting, participants must be allowed to express their reasons for interest in the campaign and to air any concerns. To stimulate constructive dialogue, the group should be presented with the assessment results of step 1, including recommendations and possible strategic options. A commitment should be sought from each participant to continue the dialogue, pursue formalization of the task force, and define roles and responsibilities of each member.

Step 3: Design the Preliminary Strategy

Once the group agrees to organize itself as a partnership around a general strategic direction, a draft marketing strategy should be prepared and a memorandum of understanding (MOU) signed by all participants committing them to specific areas of involvement and support. The draft strategy should define the objective(s) of the campaign based on the preliminary market research conducted during the assessment. Objectives should be SMART - **S**pecific, **M**easurable, **A**mbitious, **R**ealistic, and **T**ime-bound. They must clearly describe the desired outcomes of the proposed campaign. The draft strategy will also define the target audiences and will address each of the elements of the social marketing campaign, otherwise known as the "four Ps.":

Product: What is the product? What are you trying to get people to do, why is it in their interest, and how will they benefit from a change in their behaviour?

Price: What will it cost the target "consumer" in money, time or psychological terms to "buy the product" or adopt the desired behaviour? Ultimately the target group will have to be convinced that what they are being asked to do is relevant to their welfare and worth the price you are asking them to pay.

Promotion: What are the key messages for each target audience? What are the most cost-effective means of getting those messages to them - interpersonal, public relations, mass media, or advocacy? Messages need to be relevant, well focused, and, ultimately, must influence behaviour change. The messages will change over time as the target audiences evolve in their perceptions and behaviour: initially you may focus on raising awareness, later you may make a call to action.

Place: Where is the consumer expected to buy the product or act on the call to action? The campaign may have a dual focus on prevention and control, involving different places, i.e. travel and community.

Step 4: Conduct Market Research

Once the MOU has been signed and the draft marketing strategy has been developed, it will be necessary to conduct further quantitative consumer research to explore the key issues identified in the assessment. Market research will serve as a guide to all marketing decisions and be the basis for tracking the impact of the campaign. The target audience(s) must be actively involved in campaign development through market research.

A basic approach to establishing a quantitative, repeatable measure of your community's awareness of and actions regarding the targeted invasives problem is the **K**nowledge, **A**ttitudes, and **P**ractices (KAP) survey. KAP surveys query a statistically representative sample of your targeted "consumers" via telephone interviews, written questionnaires, and intercept interviews.

Step 5: Develop and Implement an Integrated Marketing Plan

The marketing plan is the blueprint for the invasive species campaign. It should include the following elements, some of which have been described above:

Situation analysis: Information compiled during the assessment, including recommendations and possible strategies.

Key issues: The assessment and market research will identify the problems and opportunities that will be encountered in designing and implementing the campaign and must be addressed in the marketing plan.

Objectives: The "SMART" objectives set by the partnership task force in the draft marketing strategy must be refined based on the additional research and partnership discussions that have taken place.

Strategies: Describe the specific strategies that will be used to achieve the group's objectives. Strategies should only focus on those whose changed behaviour will result in impact.

Advertising and public relations as strategies are excellent tools for creating awareness. Education, training and policy have longer lasting implications, which is why social marketing stresses them.

Step 6: Conduct Monitoring and Evaluation

The continued success of the marketing strategy will depend on regular monitoring and periodic evaluation. The task force must assign a project manager who has the responsibility to track progress against the marketing plan and report back to the task force on a regular basis. The project manager must also track the progress of partners in fulfilling their obligations under the MOU and reinforce their efforts as needed.

Quantitative KAP research must be repeated periodically to measure impact, guide the development of new educational and marketing materials, and guide the annual refinement of strategies. Tracking and evaluation research should use the same methodology and questionnaire used in the baseline survey.

On an annual basis the task force should conduct a review and planning process. The marketing plan should be compared with the project's success in achieving set objectives. All members are encouraged to discuss their satisfaction or frustrations. This input, combined with ongoing market research, can be the basis for a participatory planning process through which the marketing plan will be refined as needed.

2.5 Institutionalising the national commitment

The next stage in the process of making the national commitment operational is to prepare and establish a national strategy. The initial assessment is used to identify major problems, e.g. capacity, a quarantine system that focuses exclusively on agricultural pests to the neglect of natural ecosystem pests, gaps in jurisdiction, and agencies working at cross-purposes.

Ideally, a national strategy should identify, designate, or create a single lead biosecurity agency responsible for preparation and implementation of a national strategy, as recommended in the "IUCN Guidelines for the Prevention of Biodiversity Loss Caused by Alien Invasive Species". If the efforts are not focused and strengthened under a single lead agency, the resulting situation would be analogous to having separate public health agencies responsible for viral diseases, bacterial diseases, vaccine development, etc. (e.g. references in IUCN guidelines). If the responsibility for development of a national strategy cannot be assigned to a single agency, and an interagency approach is required, steps must be taken to improve co-operation and ensure a more effective way of reducing competition for funding or responsibilities and resolving conflicts of interests (Case Studies 2.9 "The Establishment of an Inter-Ministerial Committee to Control *Miconia calvescens* and Other Invasive Species in French Polynesia" and 2.10 "Co-ordination of Witchweed Eradication in the USA"). This needs to be a permanent interagency committee, which ideally should have its own dedicated staff with no other allegiances and responsibilities and its own funding. It will be necessary to have a clear definition of the roles and responsibilities of existing agencies and a formal arrangement for co-ordination of their activities in respect to alien species.

2.5.1 The national strategy

Crucial components, which need to be addressed in a national strategy, are summarized in the following:

➤ The first step is to establish the vision, goals and objectives for the invasive species strategy. This strategy must be integrated into the larger national commitment to sustainable use and an action plan for conservation of the nation's biodiversity. The ultimate goal of every initiative is the preservation or restoration of vital ecosystems and habitats with healthy, self-sustaining populations of native species. These natural ecosystems will provide important ecosystem services. Elimination of invasive organisms is one crucial tool to achieve the objective of habitat restoration, but it is not the goal for a national strategy.

➤ The strategy should also be integrated in a comprehensive framework with other national plans for public health, agriculture, conservation and other major sectors, because problems caused by invasive species cross competence borders and motivations for control of invasive pests are more often based on economic rather than ecological grounds (Case Study 2.19 "Developing a Strategy for Prevention of Invasive Alien Species Introductions into the Russian Coastal and Inland Waters") . For example a national weeds plan, such as that recently developed for Australia (Case Study 2.11 "Summary of Australia's National Weeds Strategy"), will have substantial overlap with a national strategy for invasive species.

➤ All stakeholders must be involved in the strategy from the beginning to avoid a situation in which stakeholders veto action late in the process on the grounds that they were not informed.

➤ Broad, national responsibilities for prevention, early detection, and control of invasives must be defined.

➤ A comprehensive national survey of invasives species, including their distribution, past spread and potential future dispersal, and the threats they pose, must be conducted to create a knowledge base. Capacity building in taxonomic knowledge and identification will be crucial for many countries. Research into threats posed by alien species should be linked with the survey so that information about the impact of specific species can be gathered. Research investigating the interactions between invasive species and their combined effects should also be linked with the survey. Synergistic effects between alien species and native species should be considered. Major pathways for introduction of foreign species need to be investigated. In most countries, more research will be needed on taxonomy and identification of species. There will often be a shortage of knowledge about natural distributions. For some groups, especially marine organisms, it will often be difficult even to state whether a questionable species is indigenous or introduced, i.e. their origin will be unknown (cryptogenic species). The results obtained during these studies should be disseminated to generate public awareness, and also fed into international databases to contribute to an accessible global knowledge base of invasive species (Case Study 2.12 "The Process of Determining Weeds of National Significance in Australia").

➤ While the survey gathers data about species and their impacts, priorities for action in advance of a completed national strategy should be identified based on the urgency of the problem and the values threatened, (Case Studies 2.13 "Invasive Alien Species Priorities for the South Pacific Regional Environment Programme (SPREP)" and 2.14 "Invasive Alien Species National Priorities in Mauritius"). Economic analysis can assist in prioritisation. As part of this process, urgently needed research and future research priorities should also be identified (see Section 5.1).

➤ After the results of the initial survey are finalized, a continuing process for evaluating these species and also new introductions should be established, since invasive species problems are increasing dramatically and the vectors for introduction of potential invasives change over time. Thus, continued monitoring is needed for risks posed by a changing environment, changing human practices in agriculture, forestry, etc., increasing trade, new pathways, global climatic changes such as warming due to green house effect, etc.

➤ Based on the survey, priorities for action should be identified in the areas of early detection mechanisms, prevention options, and procedures for management, control, and eradication.

➤ Invasive species and biosecurity should be a concern of all branches of government and be integrated into the missions of commerce, defence, health, agriculture, etc.

➤ A strategy needs to be defined for integrating the commitment to combat alien species into international relationships. The country has to state its approach to international obligations, e.g. under the Convention on Biological Diversity, IPPC (International Plant Protection Convention), etc., to responsibilities for neighbouring countries and nations sharing pathways (accidental export of invasive species), to participation in regional programmes, to information sharing, and to responsibilities as a source country for export of invasive species (Case Study 2.15 "Mauritius and La Réunion Co-operate to Prevent a Sugar Cane Pest Spreading").

➤ A public awareness campaign must be developed to engage the public at all stages in preventing invasions and controlling alien species (Case Study 2.16 "Priorities for Action: Hawaii"). Appropriate messages must be clearly identified to form the basis of such a campaign. Public awareness is the topic of a separate section in GISP, from which a social marketing strategy is summarized in Section 2.4.

2.5.2   Legal and institutional frameworks

It is now recognized that isolated and unilateral action by individual States can never be enough to manage the full range of activities and processes that generate invasions. Effective management requires not only national legal frameworks but also concerted bilateral, regional, or global action based on common objectives and jointly agreed international standards. Law is necessary to implement policy, set principles, rules and procedures, and provide a foundation for global, regional and national efforts.

Currently, there are more than fifty global and regional soft law instruments (agreements, codes of conduct, and technical guidance documents) dealing in one way or another with alien species. They cover terrestrial, marine, freshwater, wetlands and coastal ecosystems, as well as processes and pathways that generate introductions. A table containing a list of these international instruments is provided in the IUCN (Shine, Williams, Gündling, 2000) "A Guide to Designing Legal and Institutional Frameworks on Alien Invasive Species". This guide is also recommended as literature for more detailed information on legal issues.

International instruments are often, though not always, fairly general in character. National legislation and regulations are necessary to operationalise these instruments in national legal systems. National law, like international law, has developed by sectors over a long time scale. This sectoral approach has resulted in fragmentation, weaknesses related to coverage and terminology, and weaknesses in compliance, enforcement and remedies.

National policy makers should seek to develop a structured legal framework to address all the issues concerning alien species. As a first step, national policy makers should consider integrating alien species issues into the broader environmental and other sectoral strategic planning processes. Next, all relevant existing policies, legislation and institutions should be identified and reviewed to identify gaps, weaknesses and inconsistencies. National frameworks should be established, streamlined or strengthened to:

➤ harmonize objectives and scope,

➤ standardize terminology,

➤ implement measures to prevent unwanted introductions,

➤ support mechanisms for early warning systems,

➤ provide management measures, including the restoration of native biodiversity, and

➤ promote compliance and accountability.

In developing national law to address alien species, three approaches should be considered:

1. The first and most ambitious is to review and consolidate existing laws and regulations into a **unitary legislative framework** that covers all categories of species, sectors, ecosystems and the full range of actions to be taken.

2. The second option is to **enact one core framework** legislation that determines common essential elements, and harmonizes goals, definitions, criteria and procedures for separate sectoral laws.

3. A third option - taking a minimalist (but probably realistic) approach - is to **harmonize all relevant laws or regulations** to ensure more uniform and consistent practice.

In many countries, responsibility for alien species control is shared between various government agencies. There is often weak or no co-ordinating framework to link these agencies. Lead responsibility may be given to an existing authority, such as the environment, nature conservation, agriculture, or public health departments, or a specially established body as in New Zealand (cf. Section 2.5). Responsibility may also be shared between the relevant sectoral institutions and agencies. For this to work effectively, a co-ordination mechanism should be put in place, for example, the recently established federal Invasive Species Council in the United States.

Equally important are mechanisms to ensure co-ordination and co-operation between federal and sub-national agencies. This is particularly important for regional economic integration organizations, such as the European Union (EU) and the South African Development Co-operation (SADC), designed to promote the free movement of goods between their Member States.

Explicit objectives are necessary to provide a conceptual framework to develop the legislation, guide implementation, set priorities and build awareness. Among the main objectives are:

➤   protection of animal, plant and human health against alien pests, including pathogens and diseases; and
➤   protection of native species, including lower taxa, against contamination, hybridisation, local eradication or extinction.

The scope of national frameworks concerns two aspects: geographic and species coverage. As all parts of the national territory may be affected if an invasion takes hold, introductions should be regulated for all ecosystems and biomes - terrestrial and aquatic. Special measures are also necessary for island states or countries with islands or countries with particularly vulnerable ecosystems - such as geographically or evolutionary isolated ecosystems, including oceanic islands.

As invasive alien species are found in all taxonomic groups, including fungi, algae, higher plants, invertebrates, fish, amphibians, reptiles, birds and mammals, measures should therefore apply to all groups of species. At national level, as at international level, definitions and terminology vary widely between countries and even between sectors. General terms used by the scientific community will need to be further specified or defined in legislation to provide guidance and clarity. For legal purposes, a basic list of key terms that should be defined includes:

➤   Native species – what constitutes a native or an indigenous species?
➤   Alien species – this raises the question of "alien to what"?
➤   Invasive alien species.
➤   Threat or harm – what type or level of threats constitutes invasions?

For practical and legal purposes (see Section 3.2 for more details), a distinction must be drawn between:

➤ Intentional introductions (e.g. for agriculture, forestry, fisheries, horticulture, etc.).

➤ Intentional introductions for use in containment or captivity (e.g. aquarium, the pet trade, zoos and circuses).

➤ Unintentional introductions (e.g. through trade, tourism, travel and transport).

To the extent possible, procedures should be put in place to minimize the risk of transferring alien species at the point of origin or export. Measures of this kind are closely dependent on information exchange and co-operation between the countries concerned. In most cases inspection of commodities has to be done at the point of import or release. Border control and quarantine measures are necessary to control intentional introductions subject to prior authorization, to minimize unintentional introductions, and to detect unauthorized (illegal) introductions. Measures should also be developed for internal domestic controls, especially for:

➤ island states and states with islands;

➤ states with federal or regionalised systems of government; and

➤ regional economic integration organizations.

Introductions to protected areas, geographically and evolutionary isolated ecosystems, and other vulnerable ecosystems should be prohibited or subject to extremely strict regulation.

No intentional introduction should take place without proper authorization - usually in the form of a permit or license. Permit systems should be supported by some type of species listing technique to facilitate implementation and keep records of introductions (cf. Section 3.3). Risk analysis and environmental impact assessment (EIA) should be integral elements of the permit system (see Section 3.4).

To promote transparency and accountability, legislation should require permit decisions to be made in accordance with scientific evidence. Where a permit is granted, legislation should make it possible to attach conditions, such as preparation of a mitigation plan, monitoring procedures, containment requirements and procedures for emergency planning. Financial charges - such as a levy or deposit bond - may also be attached.

Early detection and warning systems are essential preconditions for rapid responses to new or potential invasions (cf. Section 4). The range of objectives might include requirements:

➤   to monitor the behaviour of intentionally introduced alien species to detect signs of invasiveness;

➤   to detect the presence of unintentional or unlawful introductions;

➤   to take emergency action; and

➤   to give authorities powers to take necessary and immediate action.

The risks associated with different pathways vary between countries and regions, partly in accordance with the scope and effectiveness of legal measures already in place. National measures should address **known pathways**, such as fisheries, agriculture, and horticulture industries, and monitor potential pathways. Border and quarantine controls should be designed to detect **stowaways** in consignments, containers, etc. with provisions made for post-quarantine control. Transport operations by air, sea, inland waters or land should be conducted in accordance with international and/or national standards to reduce movements of 'hitchhikers'. Special conditions should be applied for species introduced for zoos, circuses, captive breeding, pets and other contained use to reduce the risk of "**escapees**".

Environmental Impact Assessment (EIA) procedures should be modified, where necessary, to minimize the risk of introductions during large infrastructure developments. The Suez Canal, for example, now provides a permanent pathway for alien marine species to move between the Mediterranean and the Red Sea (see also Section 3.2.4 "Human-made structures which enhance spread of alien species").

Ideally, legally backed mitigation measures should have two objectives:

➤   Short- and long-term measures for eradication, containment and control of invasive aliens.

➤   Positive strategies for restoration of native biodiversity.

Alien species must have legal status compatible for mitigation programmes. In some countries all wild species may be automatically protected, including alien species. This happens where the law confers protection on biodiversity as a whole without making any distinction between alien and native species, or where it protects a higher taxon (genus) that includes an alien species.

To get round this problem, biodiversity/nature conservation legislation must exclude alien species from legal protection, and protect native species, including reintroduced species, and species that occur occasionally on the relevant territory. Mitigation measures should give powers to authorities to:

➤   regulate possession and domestic movement or trade in alien species;

➤   restrict subsequent releases;

Chapter 2
Building Strategy and Policy

- seek co-operation of owners, land owners/occupiers, and neighbouring countries;
- use cost-effective mechanisms to finance eradication, e.g. bounty systems. Techniques for eradication or control, including the use of alien biological control agents, such as ladybugs, should be subject to risk analysis/EIA and a permit from the competent authorities.

Legal frameworks should where possible support the use of incentives to promote active participation by indigenous and local communities and landowners. South Africa's Working for Water Programme provides an excellent large-scale example of such an approach.

Invasive alien species management should be seen as part of a broader suite of policies and measures to conserve biodiversity. Measures to control 'negative' biodiversity, for example clearing an area of leafy spurge, should be combined with positive incentives and strategies for restoration of degraded ecosystems and, where appropriate, re-establishment or reintroduction of native species. It is important for legal frameworks to promote a culture of civil and administrative responsibility and accountability. Approaches to promote accountability may include:

- criminal and civil liability for illegal introductions and breach of permits;
- mandatory insurance;
- deposit/performance bonds; and
- fees and charges for risk analysis and permits.

In the long-term, awareness building strategies for citizens, commercial stakeholders and administrations may make the biggest contribution to lowering the rate of introductions and effectively controlling invasions. To summarize, some of the key legal and policy principles, frameworks and tools that should be incorporated in national law include:

- strategic and long-term ecosystem management;
- co-operation - international, regional and transboundary;
- preventive and precautionary measures in control and mitigation programmes;
- cost recovery mechanisms to ensure, where possible, that the parties responsible for the introduction bear the economic burden of any necessary control measures;
- participation, including access to relevant information, by all stakeholders and relevant parties; and
- risk analysis and EIA as part of the permit procedures and mitigation programmes.

General:

http://www.issg.org/database GISP (Global Invasive Species Programme) Database and Early Warning System.

http://www.pub.whitehouse.gov/uri-res/I2R?urn:pdi://oma.eop.gov.us/1999/2/3/14.text.2 Executive Order 13112 on Invasive Species issued February 3, 1999 by US President Clinton.

http://www.invasivespecies.gov A comprehensive, online information system for the USA, developed in accordance with Executive Order 13112 on Invasive Species, and guided by the Invasive Species Council.

http://www.landcare.cri.nz/science/biosecurity/ Biosecurity and management of pests in New Zealand.

http://www.sns.dk/natur/nnis/indexuk.htm Information on who works with invasive or introduced species in the Nordic countries.

http://www.wws.princeton.edu/~ota/index.html "Harmful non-indigenous species in the United States" – a report prepared by the Office of Technology and Assessment (OTA).

http://ceris.purdue.edu/napis/pests/index.html Co-operative Agriculture Pest Survey & NAPIS' page for pest information. An extensive list of USA pests with information sheets.

http://www.nal.usda.gov/ttic/misc/picontrl.htm Management of invasive species.

http://www.iabin-us.org/biodiversity/index.htm Biological diversity information networks.

www.ramsar.org Includes guidance on designing an effective monitoring programme.

http://www.environment.gov.au/bg/invasive/ Information on invasive species in Australia.

http://www.doc.govt.nz/cons/pests/pest.htm Pest and weed fact-sheets from the Department of Conservation, New Zealand.

http://www.aphis.usda.gov/ppd/rad/webrepor.html Published APHIS (Animal and Plant Health Inspection Service) rules.

www.nbii.gov/invasives Social marketing in invasives management.

http://invasives.fws.gov/ Invasive species prevention and control programme of the U.S. Fish and Wildlife Service.

http://www.eti.uva.nl/database/WTD.html Database for world taxonomy.

Vertebrates:

http://www.nature.coe.int/CP20/tpvs65e.doc Guidelines for Eradication of Terrestrial Vertebrates: a European Contribution to the Invasive Alien Species Issue.

http://www.landcare.cri.nz/conferences/manaakiwhenua/papers/index.shtml?cowan Research paper on impacts and management of introduced vertebrates to New Zealand.

http://www.uni-rostock.de/fakult/manafak/biologie/abt/zoologie/Neozoen.html Information on non-indigenous species in Germany, with a particular focus on biology and genetics of the invaders. (In German).

Invertebrates:

http://www.hear.org/AlienSpeciesInHawaii/index.html Information about selected alien invertebrates, which are in, or that might/would be invasive or harmful if they reached, Hawaii.

http://invasivespecies.org/NANIAD.html The North American Non-Indigenous Arthropod Database contains data so far captured from diverse resources for 2,273 species of non-indigenous insects and arachnids.

http://www.uni-rostock.de/fakult/manafak/biologie/abt/zoologie/Neozoen.html Information on non-indigenous species in Germany, with a particular focus on biology and genetics of the invaders. (In German).

http://doacs.state.fl.us/~pi/fsca/exoticsinflorida.htm Lists of alien arthropod species found in Florida and some graphs illustrating the facts of invasions.

http://www.aphis.usda.gov/invasivespecies/ Within this web site are databases identifying and providing information regarding non-indigenous arthropods that have been introduced into North America, and invasive species regulated by the Animal and Plant Health Inspection Service (APHIS).

http://www.exoticforestpests.org/ The Exotic Forest Pest Information System for North America identifies exotic insects, mites and pathogens with potential to cause significant damage to North American forest resources. The database contains background information for each identified pest and is intended to serve as a resource for regulatory and forest protection agencies in North America.

Weeds:

http://www.hear.org/pier/ Pacific Island Ecosystems at Risk - Here you will find listings and descriptions of plant species, which threaten Pacific island ecosystems, particularly those of Micronesia and American Samoa. It is also available as CD by request to James Space, PIER, 11007 E. Regal Dr., Sun Lakes, AZ 85248-7919, jspace@netvalue.net

http://thomas.loc.gov/cgi-bin/query/D?c106:1:./temp/c10698Ags1: Noxious Weed Co-ordination and Plant Protection Act introduced April 29, 1999, Senate bill 910 regulates the interstate movement of weeds, including aquatic plants.

http://www.dpie.gov.au/dpie/armcanz/pubsinfo/nws/nws.html The National Weed Strategy: a strategic approach to weed problems of national significance in Australia.

http://www.agric.wa.gov.au/progserv/Plants/weeds/ Provides information on weeds and links.

http://invader.dbs.umt.edu/ US INVADERS Database System for early detection, alert, and tracking of invasive alien plants and weedy natives.

http://tncweeds.ucdavis.edu/ Database on weeds, including control methods, hosted by The Nature Conservancy.

http://www.dnr.cornell.edu/bcontrol/ Biological control of non-indigenous species.

http://ceres.ca.gov/theme/invasives.html The California Environment Resources Evaluation System (CERES): information on weeds.

http://www-dwaf.pwv.gov.za/Projects/wfw/ The Working for Water programme, Republic of South Africa – provides information on management of invasive alien plant species.

http://www.naturebureau.co.uk/pages/floraloc/floraloc.htm Flora Locale is a group under The Nature Conservation Bureau Limited representing a range of services and organizations in the UK. The use of non-local wild plant seeds, trees and shrubs for ecological restoration and schemes is discussed and projects in relation to this subject are initiated.

http://members.iinet.net.au/~weeds/linkspage.htm Weed web pages.

http://plants.ifas.ufl.edu/database.html Aquatic, Wetland and Invasive Plant Information Retrieval System (APIRS).

http://plants.usda.gov/ The PLANTS Database is a single source of standardized information about plants. This database focuses on vascular plants, mosses, liverworts, hornworts, and lichens of the U.S. and its territories. It also has a section on invasives.

http://www.agric.wa.gov.au/progserv/plants/weeds/ Extensive database and links on Australian weed species by Agriculture Western Australia.

http://www.nps.gov/plants/alien/ Provides information of weeds in the USA.

Marine focus:

http://www.marine.csiro.au/CRIMP/Toolbox.html Toolbox of eradication and control measures against marine (and some freshwater) pests.

http://www.ciesm.org/atlas/index.html CIESM's (the Mediterranean marine science research network) guides and research announcements: Guide of Mediterranean marine research institutes; Atlas of exotic species in the Mediterranean.

http://www.ku.lt/nemo/mainnemo.htm Non-indigenous species in the Baltic Sea.

http://members.aol.com/sgollasch/sgollasch/index.htm Exotics Across the Ocean: EU Concerted Action: Testing Monitoring Systems for Risk Assessment of Harmful Introductions by Ships to European Waters.

http://massbay.mit.edu/exoticspecies/index.html Information on marine bioinvasions, including pathways, prevention, and control.

http://www.com.univ-mrs.fr/basecaul Information on the seaweed *Caulerpa taxifolia*. In French.

http://www.jncc.gov.uk/marine/dns/default.htm JNCC Joint Nature Conservation Committee Directory of Introduced Species in Great Britain is a database of non-native marine species maintained by The Joint Nature Conservation Committee of Great Britain.

http://www.uscg.mil/hq/g-m/mso/ US Coast Guard Ballast Water Management Programme: Ballast water regulations; enforcement policies; exotic species information.

http://www.gmpo.gov/nonindig.html Information on non-indigenous species in the Gulf of Mexico and ballast water.

http://www.sgnis.org/ Sea Grant Non-indigenous Species website.

http://www.wsg.washington.edu/ Information on marine bioinvasions from the Washington Sea Grant.

http://www.ku.lt/nemo/species.htm Inventory of the Baltic Sea alien species of the Baltic Marine Biologists Working Group on Non-indigenous Estuarine and Marine Organisms.

http://www2.bishopmuseum.org/HBS/invert/list_home.htm Checklist of the marine invertebrates of the Hawaiian islands.

Aquatic focus:

http://nas.er.usgs.gov/ The "non-indigenous aquatic species" information resource for the United States Geological Survey.

http://www.cawthron.org.nz/index.htm Cawthron Institute: New Zealand's first private research institute, specializing in aquaculture, biosecurity, coastal & estuarine ecology, freshwater ecology and analytical laboratory services.

http://www.entryway.com/seagrant/ Sea Grant's National Aquatic Nuisance Species Clearinghouse – information on invasive non-indigenous aquatic species.

http://cce.cornell.edu/seagrant/nansc/ Sea Grants National Aquatic Nuisance Species Clearinghouse (= SGNIS). Information on aquatic invaders.

http://thomas.loc.gov/cgi-bin/query/D?c106:1:./temp/c10698Ags1 Noxious Weed Co-ordination and Plant Protection Act introduced April 29, 1999, Senate bill 910 regulates the interstate movement of weeds, including aquatic plants.

http://www.anstaskforce.gov/nanpca.htm Non-indigenous Aquatic Nuisance Prevention and Control Act introduced on November 29, 1990, and subsequently amended by the National Invasive Species Act of 1996.

http://www.great-lakes.net/envt/exotic/exotic.html The Great Lakes Information Network on exotic species.

http://plants.ifas.ufl.edu/database.html Aquatic, Wetland and Invasive Plant Information Retrieval System (APIRS)

CASE STUDY 2.1 Development of the South Pacific Regional Environment Programme (SPREP)

Prevention is the priority task in the South Pacific because control and eradication are practically extremely difficult and expensive. Specific tasks for implementing the Regional Invasive Species Strategy include:

➤ Train local experts in interception, detection and management of invasive species that threaten native biodiversity. Border control officers and conservation officers need to be able to intercept more than just species that threaten agriculture and public health.

➤ Remove pests from high profile islands and use these as advocacy models for further island restoration programmes. These model control and eradication programmes should be integrated with species or island recovery programmes and if possible with income generation such as through eco-tourism.

➤ Determine priority conservation islands which should be monitored and to which shipping and boat access should be regulated. It is impossible to protect all islands. Priority assessment needs to be brutal and based on cultural and scientific considerations.

➤ Set up contingency plans for priority islands and ensure the infrastructure is in place to deal with invasions. It may be necessary to establish experts and material resources on a sub-regional basis administered under existing regional organisations such as SPREP, SPC, and WWF etc.

➤ Commission a technical review of which invasive species occur in the South Pacific islands, which islands are at risk from what, what conservation values are currently under threat, current research and management, what practices are presenting the most threat by introducing pest species, what legislation and regulations protect islands from pest species introductions.

The first eradication demonstration project is underway in Samoa to eradicate rodents from two offshore islands. The project includes teaching pest management, eradication and monitoring skills to local staff, publicity through media releases, visits by village elders and politicians, and display boards.

A training programme has been funded for border control officers to sensitise them to invasive species, which are of particular threat to indigenous biodiversity. Part of the training will involve the trainer gathering information on the state of border control in their countries, what particular needs are and where the greatest threats are coming from. The brown tree snake will be used as a "flagship" species to focus on pathways and impact.

The technical review has been completed together with the Pacific Regional Invasive Species Strategy and should serve as the basis for funding proposals for implementation of the regional strategy and in-country projects.

Prepared by Greg Sherley, Programme Officer, Avifauna Conservation and Invasive Species; South Pacific Regional Environment Programme; PO Box 240; Apia, Samoa; E-mail: greg@sprep.org.ws

CASE STUDY 2.2   Economic Costs of Invasive Alien Species in the USA

The cost to taxpayers of introduced species in the USA was estimated, in a 1993 report of the Congressional Office of Technology Assessment, to range from hundreds of millions through billions of dollars each year. These estimates do not include effects on native ecosystems, such as extinction of native species that are of no immediate economic concern.

Best documented are costs to agriculture: about a quarter of the USA's country's agricultural gross national product is lost each year to foreign plant pests and the costs of controlling them. In the case of cotton, the total accumulated cost of the boll weevil, which arrived in the USA from Mexico in the 1890s, now exceeds 50 billion dollars. Leafy spurge, an unpalatable European plant that has invaded western rangelands, caused losses of US$110 million in 1990 alone. In eastern forests, losses to European gypsy moths in 1981 were US$764 million, while the Asian strain that has invaded the Pacific Northwest has already necessitated a US$20 million eradication campaign. To keep USA waterways clear of such plants as Sri Lankan hydrilla and Central American water hyacinth, about US$100 million is spent annually. The cost of Eurasian zebra mussels, which clog pipes in water systems such as cooling systems in power plants, is predicted to be hundreds of millions of dollars annually.

Costs of introduced pathogens and parasites to human health and the health of economically important species have never been comprehensively estimated, but must be enormous. A recent example is the Asian tiger mosquito, introduced to the USA from Japan in the mid 1980s and now spreading in many regions, breeding largely in water that collects in discarded tires. The species attacks more hosts than any other mosquito in the world, including many mammals, birds, and reptiles. It can thus vector disease organisms from one species to another, including into humans. Among these diseases are various forms of encephalitis, including the La Crosse variety, which infects chipmunks and squirrels. It can also transmit yellow fever and dengue fever. The exotic disease brucellosis, probably introduced into the USA in cattle, is now a major economic and ecological problem, for it causes miscarriages in bison and elk as well as domestic livestock.

Edited from Simberloff, D. (1996) Impacts of Introduced Species in the United States
Consequences 2(2), 13-23. *The Congressional Office of Technology Assessment*
"Harmful Non-Indigenous Species in the United States" is available at
http://www.ota.nap.edu/pdf/1993idx.html).
See also Pimentel, D.; Lach, L.; Zuniga, R.; Morrison, D. (1999) Environmental and economic
costs associated with non-indigenous species in the United States, available at
http://www.news.cornell.edu/releases/Jan99/species_costs.html

Chapter 2
Building Strategy and Policy –
Case Studies

CASE STUDY 2.3   Economic Justification for the "Working for Water" Programme in South Africa

Placing a value on ecosystem services is essential for making rational choices about competing forms of land use. In many cases, short-term economic growth and social delivery takes precedence over ecosystem conservation, so that placing a monetary value on ecosystem services is the only politically expedient way of ensuring intervention. South Africa's "Working for Water" Programme maximises and enhances sustainability of ecosystem services (chiefly delivery of water, but also cut flower trade etc.), restores and preserves biodiversity by eliminating invading alien plants; and promotes social equity through jobs, training and support for the poorest in society.

In South Africa, the introduction of hundreds of species of alien trees has led to the establishment of many populations of aggressive invaders. These trees convert species-rich vegetation to single-species stands of trees, increasing biomass and dramatically decreasing stream flow.

In the 1930s to 1950s South Africa established a series of whole-catchment experiments to assess the impacts of commercial forestry with alien species on water resources in high-rainfall areas. The results have been used to illustrate the potential impact that invasions of alien trees (as opposed to formal plantation forestry) could have on water resources, given that such invasions are comparable to afforestation.

The CSIR Division of Water, Environment and Forestry Technology mapped the extent of invasion of all important species, using local experts' knowledge, and used these data to model alien plant spread and water use. The survey covered the identity of the major invasive species, current and future possible extent of invasion, current and future impacts on water resources, and the costs of dealing with the problem.

The current invasion covers 1.7 million ha, and is estimated to be using 4400 million m^3 of water (almost 9% of the runoff of the country), based on available models of water use by trees. About 15 species (including Australian *Acacia*, *Eucalyptus* and *Hakea* species, and European and American *Pinus* and *Prosopis* species) were responsible for 90% of the problem. The cost to clear all the invasive trees would be around US$ 1.2 billion, or roughly US$ 60 million per year for the estimated 20 years that it will take. However, this could be reduced to recognize that some invasive trees do not affect watersheds, and that on-going biological control programmes will have useful impact on at least some of the major invasive trees. By excluding invasive plants that do not affect watersheds, and anticipating the benefits of biological control, clearing costs could be reduced to US$ 400 million (or US$ 20 million per year), a far more manageable target.

The monetary valuation of an ecosystem service, formalised in a benefit-cost analysis, was probably the major stimulus for the launch of the "Working for Water" programme. The fact that cutting down water-demanding alien trees is a more efficient way of delivering water from catchments than building new dams, was readily appreciated by politicians operating in a cash-strapped economy.

Abstracted from Van Wilgen, B.W.; Cowling, R.M.; Le Maitre, D.C. (1998) Ecosystem services, efficiency, sustainability and equity: South Africa's Working for Water programme. Trends in Ecology and Evolution 13, 378.

CASE STUDY 2.4   The Economic Case for Control of Leafy Spurge in North Dakota, USA

Leafy spurge (*Euphorbia esula*) is a widely established noxious weed of Eurasian origin, which can be found in every county in North Dakota. First sighted in North Dakota in 1909, it now infests over one million acres. Leafy spurge acreage was doubling every ten years for the last 30 years or more until a successful biological control programme was implemented.

A rangeland economics model was developed to estimate the economic impacts of leafy spurge infestation on both ranchers and regional economies. A leafy spurge-induced carrying capacity reduction of about 580,000 animal unit months (AUMs), or enough for 77,000 cows, reduced ranchers' annual net income by nearly US$9 million. The regional impact was about US$75 million in reduced business activity for all sectors.

An estimate was made of the regional economic impact of leafy spurge on North Dakota wildland. Wildland is land not classified as urban or built up, industrial, or agricultural, such as forest, range, or recreation areas and represents approximately 4,899,000 acres, or 10% of the state's total land area.

The biophysical impacts of leafy spurge on wildland wildlife-associated recreation, soil and water conservation, and intangible benefits resulted in direct economic impacts of US$3.6 million. Using the North Dakota 18-sector Input-Output Model, regional (North Dakota) economic impacts (direct plus secondary impacts) from leafy spurge on wildlands were estimated at over US$11 million.

Total regional economic impact (direct plus secondary impacts) from the leafy spurge infestation on wildland and rangeland in North Dakota was estimated at US$87.3 million.

Abstracted from Wallace, N.M.; Leitch; J.A.; Leistritz, F.L. (1992) Economic impact of leafy spurge on North Dakota wildland. North Dakota Farm Research 49, 9-13. and Leistritz, F.L.; Thompson, F.; Leitch, J.A. (1992) Economic impact of leafy spurge (Euphorbia esula) in North Dakota. Weed Science 40, 275-280.

CASE STUDY 2.5 Scientists Petition for Action on Green Seaweed in the USA

The following petition, signed by over 100 ecologists and exotic species research scientists and dated October 19 1998, was sent to Secretary Bruce Babbitt of the US Department of the Interior.

"An aquarium-bred clone of the green seaweed, *Caulerpa taxifolia*, has invaded the Mediterranean coasts of France and Italy and now covers thousands of acres of the coastal zone. As ecologists and exotic species research scientists, we recommend that steps be taken immediately to keep this invasive seaweed out of United States waters.

France, Spain and Australia have already banned the possession, transport or sale of this dangerous organism. However, it continues to be available for importation and sale in the United States for use in public or private aquariums. If this practice continues, it is likely only a matter of time before the Mediterranean clone of *Caulerpa taxifolia* is released and becomes established in the United States, threatening coastal waters and coral reefs from North Carolina to Florida and the Gulf of Mexico, and in southern California, Hawaii, Puerto Rico, the U.S. Virgin Islands, Guam and American Samoa. To prevent this from happening, we request that you work with the Department of Agriculture to list the Mediterranean *Caulerpa taxifolia* as a prohibited species under the Federal Noxious Weed Act, preventing its importation, possession or sale in the United States. While a native strain of *Caulerpa taxifolia* is found in tropical U.S. waters, the Mediterranean clone is a distinctly different seaweed, much larger, more aggressive, and capable of invading both tropical and cooler regions.

This invasive clone was apparently introduced into the Mediterranean Sea from the Monaco Aquarium in 1984. It covered roughly one square yard in 1984, spread to over 2 acres by 1989, and now covers over 10,000 acres extending from the shore to depths of over 250 feet. It grows on both rocky and sandy bottoms, from protected bays to exposed capes, and attains great densities, forming monoculture stands whose impact has been compared to unrolling a carpet of astroturf across the bottom of the sea. In these regions it causes ecological and economic devastation by overgrowing and eliminating native seaweeds, seagrasses and invertebrates (such as corals, sea-fans and sponges). It has harmed tourism, destroyed recreational diving and created a costly impediment to commercial fishing.

Allowing the release of this organism into the Mediterranean was an act of environmental mismanagement that threatens catastrophic changes and the loss of biological diversity in one of the world's most valued marine ecosystems. We believe that allowing its importation and sale in the United States is equally ill advised.

We further ask that you initiate a review of federal policies and practices regarding the importation and sale of non-indigenous marine and freshwater organisms by the aquarium trade. These practices generally take a "dirty list" approach, in which certain unacceptable species are prohibited and all unlisted species are freely imported and sold. It is in part this dirty list regulatory approach that has allowed the continued importation of the Mediterranean *Caulerpa taxifolia* clone and other potentially harmful organisms, despite clear evidence in some cases of substantial damage in other parts of the world.

Recent, well-documented cases of substantial economic and environmental harm caused by non-indigenous aquatic organisms demonstrate that it is time to move to a "clean list" approach, in which the United States would import only those organisms which evidence indicates will not be harmful. At stake is nothing less than the health of our commercial and recreational fisheries, the growing aquaculture and mariculture industries, and the rivers, lakes and coastal waters of our nation."

See http://www.mcbi.org/caulerpa/babbitt.html.

Despite subsequent efforts to prevent introduction of this invasive seaweed, the inevitable bad news hit the conservation community in 2000. Immediate eradication efforts have been undertaken, but the success is not guaranteed. News release (edited):

SAN DIEGO, California – A mutant algae, *Caulerpa taxifolia*, responsible for killing marine life throughout the Mediterranean has now invaded the seas off San Diego. Divers monitoring the growth of eelgrass, transplanted to restore marine habitat off of Carlsbad, California, about 20 miles north of San Diego, discovered the algae on 12 June 2000 in a lagoon near the Cabrillo Power Plant I. This is the first time the algae has been discovered anywhere along the coasts of North and South America. So far, it has been detected only in the Agua Hedionda lagoon, where the largest patch measures 60 feet by 30 feet. Nine smaller patches have also been discovered. Scientists are moving quickly to destroy the algae.

CASE STUDY 2.6   Learning from Others' Experience: The Case of *Miconia calvescens*

In April 1993, Jean-Yves Meyer, a French researcher conducting studies on bush currant (*Miconia calvescens*) in Tahiti, visited Hawaii to see *M. calvescens* populations on Maui and Hawaii, and made contacts with numerous agencies and individuals. In June 1994, Hawaii-based Arthur Medeiros of the U.S. National Biological Service (now U.S. Geological Survey: http://www.nbs.gov/) was sent to Tahiti and, assisted by Meyer, obtained good photographic documentation of the situation in Tahiti and French Polynesia. This photographic documentation of potential damage by *M. calvescens* has proved invaluable in convincing doubters in Hawaii of the need for prompt action to manage this weed.

Edited from "Miconia calvescens in Hawaii: a summary" prepared by L. Loope (March 1996), with extensive borrowings from manuscripts by Medeiros, Loope and Conant and by Conant, Medeiros and Loope, and posted on the internet through:
http://www.hear.org/MiconiaInHawaii/index.html.

CASE STUDY 2.7   The Dirty Dozen - America's Least Wanted Alien Species

One public awareness tool that is in use in North America is the Nature Conservancy's "Dirty Dozen" of unwanted alien species. Pesticide Action Network's "Dirty Dozen pesticides" (e.g. http://pnews.org/art/2art/bantrade.html) used a similar approach to identify the worst chemical pesticides in order to promote their concerns, and apply pressure on those producing, distributing and using these pesticides. The Nature Conservancy (http://www.tnc.org/) introduces their "Dirty Dozen" thus:

"The "Dirty Dozen" is a rogues' gallery representing some of America's least wanted alien species. Although these 12 intruders differ from each other in many ways, all share a common trait: they spell trouble for our native species and ecosystems. The following portfolio of pests was chosen to illustrate the breadth of problems that our native biodiversity and natural lands face from the onslaught of introduced species. Many others could have been selected-species that also are despoiling our ecosystems and imperilling our native plants and animals. The "Dirty Dozen" were chosen for this dubious distinction because they exemplify the worst of a bad lot. The species profiled here depict an array of different organisms (plants and animals), a variety of ecological systems (terrestrial, freshwater, and marine), and a wide geographical range-from Hawaii to Florida, and Maine to California."

- flathead catfish
- purple loosestrife
- rosy wolfsnail
- tamarisk
- Chinese tallow
- hydrilla
- green crab
- brown tree snake
- miconia
- balsam woolly adelgid
- zebra mussel
- leafy spurge.

CASE STUDY 2.8 Developing a Strategy for Improving Hawaii's Protection Against Harmful Alien Species

The current effort to strengthen Hawaii's quarantine systems has developed in three stages. During 1991 and 1992, two non-governmental organizations (The Nature Conservancy of Hawaii and the Natural Resources Defence Council) prepared a report entitled The Alien Pest Species Invasion in Hawaii: Background Study and Recommendations for Interagency Planning. This report describes the roles, legal mandates, and resources of each agency or organization involved in preventing pests from becoming established in Hawaii or in controlling established pests. It identifies at a general level the major problems in the current system, and recommends a process for developing plans to resolve these problems. The report highlighted two major needs above all others: a comprehensive pest management strategy linking the various players in a co-ordinated system, and strong public support and high-level political leadership as essential ingredients for success.

The 1992 background report set the stage for multi-agency development of an Alien Species Action Plan in 1993-94. This effort involved over 80 individuals from more than 40 government, non-profit, and private agencies, organizations, and businesses, who worked in professionally facilitated topic groups to prepare the plan. These topic groups submitted 34 more or less specific proposals for improvements to an oversight committee made up of leaders of key agencies and organizations. This committee then prepared the final plan, described as its commitment to "a first set of actions...to improve pest prevention and control for Hawaii." The Oversight Committee's first action was to re-form itself as a permanent Co-ordinating Group on Alien Pest Species (CGAPS). CGAPS' most important feature is the broad set of interests it represents beyond the expected state and federal quarantine agencies. These include the state transportation and health departments, the Hawaii Visitors Bureau, the Hawaii Farm Bureau Federation which also represents horticultural interests, the U.S. Postal Service, the military, and state, federal, and non-profit biodiversity conservation agencies. The group is "held together by the voluntary efforts and enlightened self-interest of its members rather than by any formal authority," although formal agreements may be desirable for certain joint programmes. Its purpose is "to expedite communications, problem-solving, and decision-making for more effective implementation of pest prevention and control work." The group is administered by the Hawaii Department of Agriculture, with additional staff support from The Nature Conservancy, and has held half-day, quarterly meetings since January 1995.

Edited from http://www.hear.org/AlienSpeciesInHawaii/articles/norway.htm "An alliance of biodiversity, agriculture, health, and business interests for improved alien species management in Hawaii" presented at the Norway/UN Conference on Alien Species, July 1-5, 1996, by Alan Holt, The Nature Conservancy of Hawaii, 1116 Smith Street, Suite 201, Honolulu, Hawaii 96817.

CASE STUDY 2.9   The Establishment of an Inter-Ministerial Committee to Control *Miconia calvescens* and Other Invasive Species in French Polynesia

In August 1997, the Délégation à la Recherche (= Department of Research, under the authority of the Ministry of Health and Research, Government of French Polynesia) organized the "First Regional Conference on Miconia Control" in Papeete, Tahiti, on the initiative of the scientist in charge of the *M. calvescens* control and research programme since 1992 (Dr. Jean-Yves Meyer). Biologists and managers from Australia, Fiji, France, French Polynesia and Hawaii (USA) attended this free public conference whose main goal was to assess the past and current efforts to control *M. calvescens*, an alien tree considered as the most aggressive invader in the native wet forests of Hawaii and French Polynesia. During the meeting final discussion, the need for strengthened collaboration between governmental departments in French Polynesia was emphasized, and the creation of an inter-ministerial committee in charge of the co-ordination of the *M. calvescens* control efforts was proposed.

An "Inter-Ministerial Technical Committee to Control Miconia and Other Invasive Plant Species Threatening the Biodiversity of French Polynesia" was officially created one year later (Decree N°1151 CM, voted by the Council of Ministers in August 1998). This important institutional step for the management of biological invasions in French Polynesia was made possible thanks to existing legislation on nature protection in this French overseas territory (Law N°95-257 AT, voted by the Territorial Assembly in December 1995). The committee, chaired by the Minister of Environment (or his representative) is assisted by the scientist in charge of the research programme on invasive plants in French Polynesia. It is composed of the governmental agencies which are actively or potentially involved in the prevention and the control of introduced plant species: la Délégation à l'Environnement (Dept. of Environment), la Délégation à la Recherche (Dept. of Research), le Service du Développement Rural (Dept. of Agriculture), la Direction de l'Equipement (Dept. of Equipment); le Service de l'Administration et du Développement des Archipels (Dept. of Administration and Development of the Archipelagos) and le Service du Tourisme (Dept. of Tourism). The committee members (head of each department or his/her representative) meet once a month, and can invite other non-governmental participants chosen because of their relevance to the action plans (e.g. research scientists, school directors, French Army representatives, nature protection groups).

The main goals of the committee are: (1) to define short- and long-term control/management strategies; (2) to find suitable human and material means, including adequate funding; (3) to set up priorities concerning public information, education, research and regulation texts. The committee has also started to address alien animal species. Action plans defined by the committee are submitted for approval to the Council of Ministers. Some relevant results of this committee include:

➤ A list of 13 dominant invasive plant species legally declared a threat to the biodiversity of French Polynesia (Decree N°244 CM, February 1998). New introductions, propagation, cultivation, and transportation between islands are strictly forbidden, and destruction is authorized. A leaflet describing these species, and including other potential plant invaders, was prepared in 1999.

➤ The organization and funding of one-week *M. calvescens* control campaigns on the island of Raiatea in June 1999 and in June 2000, with the participation of 90 soldiers of the French Army led on the field by the forestry section of the Department of Agriculture.

The inter-ministerial committee has managed to bring different agencies together for joint action to prevent, contain and eradicate plant (and animal) invasions, thus enhancing considerably the conservation efforts in French Polynesia.

Prepared by Jean-Yves Meyer, Délégation à la Recherche, B.P. 20981 Papeete, Tahiti, French Polynesia. E-mail Jean-Yves.Meyer@sante.gov.pf

CASE STUDY 2.10   Co-ordination of Witchweed Eradication in the USA

The Asian and African witchweed (*Striga asiatica*) grows parasitically on the roots of members of the Poaceae, especially maize and sorghum causing significant crop losses, but also on rice, millet, sugar cane and grasses. It was first found in the USA in 1956, and the infestation was ultimately found on 200,000 hectares spread over an area of 20,000 km^2 of eastern North and South Carolina.

When this infestation was discovered, its agronomic significance was made clear to US Department of Agriculture (USDA) officials and congress by Dr. A.R. Saunders of South Africa, an expert on this species, who was visiting the USA at that time, and witchweed was declared a national threat to USA agriculture. Federal and state quarantines were imposed on the infested area and a federally funded eradication effort was initiated.

One of the first recognized needs for successful eradication was research to develop eradication methods. A research station and test farms were established and a scientific team assembled. The herbicide 2,4-D applied throughout the growing season, with high clearance spray equipment, was found to be quite effective in maize but was not adequate for the eradication effort. Other herbicides and control measures were developed to control grassy host weeds in cotton, soybeans, horticultural crops and all other situations were the infestation occurred. Major improvements were made in equipment for more effective application of herbicides to all infested areas. A key breakthrough in eradication was the discovery that ethylene gas could cause suicidal germination of witchweed seeds in soil, and methods and technology were developed to exploit this.

The witchweed eradication programme was a co-ordinated effort led by USDA involving other federal and state agencies, agribusiness and the general public. The Animal and Plant Health Inspection Service (APHIS) of USDA was responsible for developing and conducting control activities in co-operation with North Carolina State University. The extension service provided education to farmers and landowners. The North Carolina Department of Agriculture was involved in regulatory activities. The Farm Bureau and other agricultural organizations helped secure funding. Clubs such as 4-H assisted in education and getting people to report suspected infestations. The ASCS (now Farm Service) assisted in mapping and identifying property owners. Newspapers and other media provided detailed and ongoing coverage of the problem and the eradication effort. This combined effort led everyone to recognize witchweed as everyone's problem.

Over the 45 years of the eradication programme, more than US$ 250 million has been spent. This is a small cost compared to the US$ 25 billion per year value of crops threatened by this parasitic weed. By the end of 1999, the eradication effort had reduced the witchweed-infested area to about 2,800 hectares of very light occurrences. The programme is expected to progress until finally eradication is achieved. The size of the witchweed infestation, the complexities of eradication and the time required to eradicate would normally discourage a country from starting such a major project, but the long-term benefits to agricultural productivity of the USA make this a wise investment.

Prepared by Robert E. Eplee, USDA (retired). See also:
*Sand, P.F.; Eplee, R.E.; Westbrooks, R.G. (1990) Witchweed Research and Control in the United States, Monograph Series of the Weed Science Society of America **5**, 154 pp.*
*Eplee, R. E. (1992) Witchweed (Striga asiatica): an overview of management strategies in the USA. Crop Protection **11**, 3-7.*

CASE STUDY 2.11   Summary of Australia's National Weeds Strategy

Goal 1: To prevent the development of new weed problems

<u>Objective</u>: To prevent the introduction of new plant species with weed potential

➤ Strengthen import entry protocols for assessing the weed potential of all proposed new plant imports.

➤ Initiate community education programmes to increase awareness of the use of native plant species in preference to import of some new plants

<u>Objective</u>: To ensure early detection of, and rapid action against, new weed problems

➤ Initiate community education programmes to increase awareness and facilitate early warning of new weed occurrences

➤ Co-ordinate plant identification and reporting mechanisms involving State herbaria and other expert bodies

➤ Develop guidelines for assessing the weed risk of plant material being used for breeding or selection trials prior to its release for commercial use

➤ Develop a contingency plan, identifying key groups, reporting procedures and a funding mechanism

<u>Objective</u>: To reduce weed spread to new areas within Australia

➤ Provide guidelines to States and Territories to ensure appropriate consistency in weeds legislation

➤ Facilitate adoption of the guidelines

➤ Encourage State, Territory and Local Governments to develop contingency plans for action against new weed infestations

➤ Establish effective procedures for restricting the spread of new weeds within Australia, for example, hygiene practices, machinery cleaning codes of practice, controls on nursery plant and seed sales

➤ Educate landowners, land users, industry and the general public in procedures to restrict the spread of weeds

Goal 2: To reduce the impact of existing weed problems of national significance

<u>Objective</u>: To facilitate the identification and consideration of weed problems of national significance

➤ Develop guidelines and a procedure to establish when weed problems are of national significance

➤ Strengthen existing weeds specialist networks to ensure that information to assess weed problems is readily accessible

➤ Establish procedures for assessing the relative priority of weed issues of national significance

continued...

Objective: To deal with established weed problems of national significance through integrated and cost effective weed management

➤ Develop mechanisms for assembling the information required to develop management strategies for the problems

➤ Establish procedures for developing cost efficient and effective management plans

➤ Establish procedures for implementing, monitoring and evaluating the management plans

➤ Provide guidelines to ensure that wherever possible, the Landcare approach (co-ordinated community action) be considered the appropriate delivery mechanism for much of the on ground action on weed issues implemented under this Strategy.

Goal 3: To provide the framework and capacity for ongoing management of weed problems of national significance

Objective: To strengthen the national research, education and training capacity to ensure ongoing cost effective, efficient and sustainable weed management

➤ Integrate and co-ordinate weed research, education and training programmes throughout Australia

➤ Facilitate and co-ordinate the delivery of training and awareness programmes in integrated weed management for land owners/managers and other on ground resource users

➤ Encourage tertiary institutions to emphasize, in weed science courses, the need to adopt integrated weed management practices across all ecosystems

Objective: To encourage the development of strategic plans for weed management at all levels

➤ Promote the benefits of developing complementary strategic plans for weed management at the State, regional, catchment, locality and property scale

Objective: To establish institutional arrangements to ensure ongoing management of weed problems of national significance

➤ Ministerial Councils will nominate an appropriate body to co-ordinate cross sectional issues and actions relating to weeds

➤ Establish a position of Co-ordinator: National Weeds Strategy

➤ Develop triennial plans for action on weed issues of national significance

➤ Establish a mechanism for resolving sectoral conflicts on weed issues

Extracted from http://www.weeds.org.au/nws-doc.htm

CASE STUDY 2.12   The Process of Determining Weeds of National Significance in Australia

Over the past decade there has been a developing awareness of Australia's weed threat that achieved formal recognition with the launch of the National Weeds Strategy in mid 1997 (see Case Study 2.11 "Summary of Australia's National Weeds Strategy"). A central component of the strategy is the identification of 20 Weeds of National Significance and the resultant co-ordinated actions across all States and Territories. The development of an assessment process, nomination of candidate species, assessment and resultant ranking of species, through to final endorsement by Ministerial Councils had no precedent and took two years to complete.

Seventy-four weed species were nominated by the State and Territories to be assessed against the criteria for Weeds of National Significance and represent, in their opinions, the most serious weed problems in Australia. Four major criteria were used:

➤ invasiveness,
➤ impacts,
➤ potential for spread, and
➤ socio-economic and environmental values.

Five main data sources were used for the Weeds of National Significance analysis:

➤ an invasiveness and impacts questionnaire was submitted to three expert panels covering weeds for temperate, sub-tropical and tropical environments;

➤ observed distribution and density for each weed provided by State and Territory agencies and sourced from the literature. This data and published literature was used to predict potential distribution of weeds using climatic modelling;

➤ economic information on the cost of control for agricultural and forestry weeds provided by State and Territory agencies;

➤ environmental information on the number of threatened species, communities and IBRA regions provided by State and Territory agencies and the monoculture potential of a weed from the expert panels;

➤ a qualitative assessment by the expert panels of social impacts caused by a weed (not examined by other data sources).

The NWSEC undertook an extensive analysis of the data to investigate the impact of numerous weighting schemes as they affected the ranking of the weeds. The lack of appropriate national datasets, number of species assessed, variability of some of the data resulting from different methods of recording made the analysis more difficult. This necessitated that substantial verification and standardization be applied to the data. Despite these difficulties, the datasets are considered credible, being the best data available on which to make the Weeds of National Significance decision.

The method used is not a scientific process, but an attempt to draw together meaningful indicators (where few national datasets exist) and combine them into a form that provides an objective, transparent and defensible ranking system for weeds. The relativity of the scores for individual species among a group of species is more important than the definitive scores for ranking purposes. The NWSEC is of the opinion that the data sources and analysis undertaken to determine the 20 Weeds of National Significance is the most comprehensive available and is sufficiently objective as to be clearly defensible.

Edited from http://www.weeds.org.au/nws-doc.htm

CASE STUDY 2.13   Invasive Alien Species Priorities for the South Pacific Regional Environment Programme (SPREP)

Since 1985 the Pacific countries have run conferences every four years whose main purpose has been to table the needs of conservation in the region. The 1989 South Pacific Parks and Reserves Conference (later known as the South Pacific Protected Areas Conference) resolved that the region needed an invasive species programme that would be best administered by the South Pacific Regional Environment Programme (SPREP). The conference recognized that the region's countries had suffered some of the greatest loss in biodiversity in the world compared to the number of species before man colonized the Pacific islands. The conference also recognized that today the greatest threat to most remaining native species was from invasions by alien species.

In 1991 a proposal to implement this resolution was written by New Zealand and SPREP colleagues and was finally tabled with the New Zealand Ministry of Foreign Affairs and Trade in about 1994. In 1997 funding was approved for half a fulltime position on terrestrial invasive species with the other half tackling a related problem – the conservation of birds.

Since the position was filled in September 1998 invasive species and bird conservation have been used to support each other. Thus most in-country bird species recovery programmes involve controlling introduced pests such as rats. In this way, the high profile of critically endangered birds has raised awareness of the threat of invasive species. The priorities of the programme are set out in Case Study 2.1 "Invasive Alien Species Priorities for the South Pacific Regional Environment Programme (SPREP)".

Prepared by Greg Sherley, Programme Officer, Avifauna Conservation and Invasive Species; South Pacific Regional Environment Programme; PO Box 240; Apia, Samoa; E-mail: greg@sprep.org.ws

CASE STUDY 2.14 Invasive Alien Species National Priorities in Mauritius

The following selected and slightly edited priority recommendations referring to invasive alien species management were drawn up at the Workshop on the Restoration of Highly Degraded and Threatened Native Forests in Mauritius, September 1997.

1. Deer and pigs. Using the provisions of the relevant legislation, ensure reduction to near-zero numbers, or elimination, of deer and pigs within the Park using traps, rifles, dogs and other appropriate techniques.

2. Monkeys (*Macaca* spp). A much more concerted effort must be made to eliminate monkeys from Conservation Management Areas (CMA) within the National Park. In addition to trapping, shooting and poisoning should be tested and appropriate incentives provided to appropriate responsible groups.

3. Fencing. Advantage should be taken of recent progress in Australia and New Zealand with fencing technology. In particular, tests should be conducted to develop solar-powered electrified fences suitable for excluding monkeys, deer and pigs from CMAs or other selected areas.

4. Cats. Investigate the Australian conservation work on feral cat management with a view to applying the same techniques to Mauritius.

5. Mongooses. Continue studies to gain greater understanding of their impacts and behaviour, and the development of control methods.

6. Eradicate shrews, wolf snakes and giant African snails from Ile aux Aigrettes.

7. Carry out studies to measure the effects of tenrecs and shrews on native plants and animals.

8. Other problem aliens, including exotic birds, the wolf snake, exotic lizards, exotic snails and toads. Studies are needed of the effects of these animals on native fauna and more general interactions within the Mauritius ecosystem.

9. Test a greater range of herbicides and methods of application on major weeds, for example test new herbicide application techniques such as basal bark painting, and cut and treat stump.

10. Establish a Weed and Predator Management Emergency Fund for unforeseen circumstances (Cyclone, Fire) and New Species Invasions.

11. Mauritius must become involved in the biological control programmes on: *Rubus alceifolius* and *Ligustrum robustum* due to start for La Réunion, and *Psidium cattleianum* small-scale research being undertaken in Brazil / Hawaii / UK. Involvement in these programmes must be from the beginning, to ensure that Mauritian flora is considered in any agent testing. Biological control projects should be Mascarene-wide, i.e. regional. Financial implications need consideration.

12. Consider an international project on *Psidium* to increase resources. For example, the genus is invasive in Mauritius, Réunion, Seychelles, Comoros, Galapagos, Hawaii, Norfolk, Madeira, and French Polynesia.

13. Manage conflicts of interest early, particularly with respect to *Psidium cattleianum* where local use for fruit and pole cutting will have to be considered. Education of the public and authorities concerning biological control is essential and should start now.

14. Trial on removing only portions of weed infestation and replanting natives: try only one removal versus several; measure regeneration within weedy plots with/without weed removal treatments.

15. Fill gaps in CMAs created as a result of weeding out pioneer species.

CASE STUDY 2.15   Mauritius and La Réunion Co-operate to Prevent a Sugar Cane Pest Spreading

The white grub or ver blanc, *Hoplochelus marginalis* is a polyphagous beetle whose root-boring larvae can cause huge losses to sugar cane crops. It is indigenous to Madagascar, and does not occur naturally in the Indian Ocean islands.

In 1981, the first *H. marginalis* damage was noted close to Gillot, the port and airport area of La Réunion, 760 km east of Madagascar and 150 km to the west of Mauritius, and it has since spread to all suitable areas of La Réunion. Prompt action was taken to prevent the importation of *H. marginalis* into Mauritius, which is very heavily dependent on sugar cane. The 'Plant Introduction and Quarantine Standing Committee' was immediately set up, composed of members of the Mauritius Sugar Industry Research Institute (MSIRI), Mauritian Ministry of Agriculture and the University of Mauritius. All measures formulated were carried out in consultation with CIRAD (Centre de Coopération Internationale en Recherche Agronomique pour le Développement) and the Direction Départmentale de l'Agriculture et des Forêts (DAF) of La Réunion. Quarantine measures formulated included:

➤ Changes in flight and boat departure times. During the summer the beetle actively flies around dusk and is attracted to light. No planes are allowed to take off from La Réunion to Mauritius between 18.30 and dawn. Similar restrictions apply to shipping and if ships have to stay overnight in La Réunion they must keep their lights off. All boats and planes are sprayed as necessary.

➤ Regular inspections using light traps around high-risk areas in Mauritius (around the airport and port areas).

➤ Regular spraying around the Mauritius airport region.

Regular meetings are held between specialists in Mauritius and La Réunion to assess the situation, and a 'Protocole d'accord' has been signed by Mauritius and La Réunion to ensure that the above measures are applied.

A vital part of the strategy of control has been the reduction in the population densities of *H. marginalis* especially around the port of La Réunion. This has been achieved by the use of the fungal pathogen *Beauveria brongnartii*. Adult beetles are dipped in a fungal suspension and released to spread the infection. This has resulted in relatively low populations of *H. marginalis*, which has considerably lessened the chances of its accidental introduction into Mauritius.

Public awareness campaigns have also been intense and sustained. Posters are to be seen at the airports in Mauritius and La Réunion. There are very few people in Mauritius today who are not aware of the menace posed by the 'Ver Blanc'.

Prepared by John Mauremootoo, Plant Conservation Manager, Mauritian Wildlife Foundation, Fourth Floor, Ken Lee Building, Port Louis, Mauritius; e-mail mwfexec@intnet.mu

CASE STUDY 2.16   Priorities for Action: Hawaii

Case Study 2.8 "Developing a Strategy for Improving Hawaii's Protection against Harmful Alien Species " describes how the Co-ordinating Group on Alien Pest Species (CGAPS) was established in Hawaii. Starting a public awareness campaign brought CGAPS' members face to face with the obvious question: "What specifically do we want the public and our elected officials to do once they become aware of the magnitude of the alien species problem?" Like any highly complex problem, some parts of the solution are apparent and relatively simple while others are not yet clear. CGAPS regards the following as the areas most in need of improvement:

Self-sustaining public education programme. Hawaii's greatest opportunity for improved pest prevention lies in educating the public. CGAPS' goal is to establish a dedicated funding source for continuous, high-quality public education messages delivered through a wide range of vehicles (e.g. tourist information, in-flight materials, information boards in baggage claim areas, school curricula, etc.).

Developing the ability to inspect all pest pathways. A large proportion of the total passenger, cargo, mail and other traffic entering Hawaii is currently not inspected, including materials known to be significant sources of alien species.

Systems to monitor total pest traffic. The quarantine inspection agencies can not monitor the total pest traffic through a particular pathway as a gauge on the effectiveness of quarantine programmes, and they do not currently have the resources to investigate newly detected pests to determine how they entered the state in order to detect leaks in the prevention system.

Technical support and timely processing of import permit review decisions. Although the Hawaii Department of Agriculture has the most comprehensive regulations in the USA for review of animal, plant, and microorganism imports, the expert committees that recommend permit decisions to the Board of Agriculture lack ready access to information relevant to assessing the subject taxon's disruptive potential. Decision-making is an inconsistent and time-consuming process because of this and the processing time for many permits is over 12 months.

Early detection and eradication of new pest infestations. This is the most neglected phase of the invasion process, as virtually all pest management effort is directed at port-of-entry inspections and the control of widespread pests. A database of information from diverse sources is being organized on known pests. This information can be used to identify infestations that may be vulnerable to containment or eradication on a statewide, whole-island, or regional scale.

Further, eradication of incipient invasions requires better training for managers in pest control strategies to maximize the chances for success. Too often, the initial treatment of an infestation is intense but short-lived, and not followed up to ensure complete eradication. Nor are exhaustive monitoring and activities to prevent re-infestation or spread to other sites always put in place. A commitment to better training and planning should improve the rate at which these projects succeed.

Edited from http://www.hear.org/AlienSpeciesInHawaii/articles/norway.htm *"An alliance of biodiversity, agriculture, health, and business interests for improved alien species management in Hawaii" presented at the Norway/UN Conference on Alien Species, July 1-5, 1996, by Alan Holt, The Nature Conservancy of Hawaii, 1116 Smith Street, Suite 201, Honolulu, Hawaii 96817.*

CASE STUDY 2.17 The Human Dimensions of Invasive Alien Species

The non-technical aspects of alien invasions often determine the success or failure of efforts to limit their impacts and protect biodiversity. In September 2000, Jeff McNeely, Chief Scientist of IUCN, chaired a GISP-sponsored workshop to address these human dimensions. The outline below is a portion of his summary of the participants' discussions. These aspects of invasions span all areas of human experience and range across a wide field of inquiry. Most are little studied.

Historical Dimensions, e.g.:
How we have thought and behaved in the past

Political Dimensions, e.g.:
Who has a stake and who holds power; who decides, with what discretion
"Who plays what tune and who dances to it"
What priorities we set; what boundaries we maintain
How we mobilize support

Legal Dimensions, e.g.:
Who has rights and who assigns them
What our laws say and what they don't
Which laws we enforce and which we break

Economic Dimensions, e.g.:
How we spend money and how fast; what we count; what we tax
Who we employ and whose development matters

Sociological Dimensions, e.g.:
Where we go and where we leave
How we compete and disagree; when we compromise
How we integrate our efforts; how we involve others
How we intervene; how we strike balances
Which opportunities we seize and which we lose
Which problems we anticipate and which we ignore

Cultural Dimensions, e.g.:
What values our organizations promote and what reputations they have
What songs and poems and books we write; what arts and crafts we make
What we find beautiful

Linguistic Dimensions, e.g.:
What words we use; what stories we tell; who tells them

Psychological and Ethical Dimensions, e.g.:
What we perceive and sense and feel; what motivates us and how we behave
What identity we choose; what entitlements we claim
Which attachments we make and what losses we suffer

Educational Dimensions, e.g.:
What we know and who knows it
Who we train and at what level
How we communicate; how we evaluate what we do

Philosophical Dimensions, e.g.:
What we value and how that changes with time and place
What purposes we have and what vision we hold; what we owe the future

Spiritual and Religious Dimensions, e.g.:
What we hold sacred; what rituals we practice
What miracles we long for; how we keep our spirits inviolate

Prepared by Phyllis Windle, Senior Scientist, Union of Concerned Scientists, Washington, DC, USA, pwindle@ucsusa.org. For further insight see documents prepared by this GISP Human Dimensions group (check the GISP website http://jasper.stanford.edu/gisp/ for details as they become available).

CASE STUDY 2.18 A National Assessment of Invasives: the U.S. Office of Technology Assessment Report

In 1990, the U.S. Congress was worried about alien zebra mussels in the Great Lakes. It turned to the Office of Technology Assessment (OTA), one of its research agencies, to determine whether zebra mussel was just the tip of the invasion iceberg. Specifically, Congress asked OTA to determine the economic and environmental impacts of all the nation's invaders; how effective federal policies were; what role state laws played; and the relationship between invaders and genetically modified organisms. In 1993, OTA published the results of its research: *Harmful Non-Indigenous Species in the United States*, a 400-page report.

The report was written in-house, by a four-person staff of three biologists and an environmental attorney - three hired temporarily for the study. A couple hundred experts supplemented the staff's work. For instance, a 22-person advisory panel met several times to oversee the work. Eight federal officials linked OTA with executive branch agencies. One workshop was held on decision-making methods. In addition OTA commissioned around 20 reports by academic and other experts:

➤ 6 on the pathways and consequences of introductions of various taxonomic groups, i.e. non-indigenous vertebrates; fishes; insects and arachnids; plants; freshwater, terrestrial and estuarine molluscs; and plant pathogens.

➤ 3 on decision-making models, including economic ones.

➤ 3 on policies of the major federal agencies.

➤ 3 on states' situations, i.e., on Hawaii, Florida, and a survey of state fish and wildlife laws.

➤ several on special topics, e.g., bioengineering, ecological restoration, and public education.

These papers were tailored to answer specific questions and each was peer reviewed for accuracy. Drafts of the final report were also reviewed extensively.

When the report was published, the United States had, for the first time, a national assessment that provided information regardless of taxonomic group, economic sector, and government agency. A number of its features have proven especially significant: estimates of the total number of non-indigenous species in the U.S.; their probable economic costs; a list of foreign species first detected between 1980-1993; a compilation of the responsibilities of the 20 or so relevant federal agencies; and not just detailed summaries of state fish and wildlife laws but also managers' assessments of their adequacy.

The study was neither cheap (estimated cost $700,000) nor quick - which helped ensure its thoroughness. It laid the foundation on which subsequent, more detailed, work has built. Many call it "the Bible" on U.S. invaders.

Prepared by Phyllis Windle, Senior Scientist, Union of Concerned Scientists, Washington, DC, USA, who directed the OTA study. The report can be located by date and title after selecting "OTA Publications" on the Internet at http://www.wws.princeton.edu/~ota/index.html

CASE STUDY 2.19   Developing a Strategy for Prevention of Invasive Alien Species Introductions into the Russian Coastal and Inland Waters

The invasions of invasive alien species, associated with the release of ballast waters of ships and the headlong practice of deliberate introductions, have caused significant losses in biodiversity and economy in the Former Soviet Union countries.

Growing concern in the Russian scientific community about the consequences of alien aquatic species introductions in 1998 resulted in the establishment of the Group of Aquatic Alien Species (GAAS) at the Zoological Institute of the Russian Academy of Sciences, supported by the government's "Biodiversity" programme.   Dissemination of information on aquatic invasive species for legislators, decision-makers and the general public in Russia is one of the main GAAS goals.

During 1998-1999 the GAAS scientists started publishing information on the problem, including the official report to the Russian authorities on invasive alien species introductions into the Gulf of Finland area. As a result, in 1999 the Working Group on "Biological Pollution" of the Gulf of Finland Basin has been established at the St. Petersburg Scientific Centre of the Russian Academy of Sciences. Development of a regional management plan for control and prevention of alien and pathogenic organisms in aquatic ecosystems in the Gulf of Finland Basin is the main goal of the Working Group. At present the Working Group is focusing on the transfer of scientific information on aquatic invasive species to the level of decision-makers and legislators.

In 2000 the GAAS scientists prepared a report entitled *Consequences of Alien Species Introductions and Need of Preventive Actions*, which has been published in the proceedings of the first national seminar on introduced species in the European seas in Russia. This report highlights the needs of the national management plan for control and prevention of invasive alien species introductions in the Russian coastal and inland waters.

Edited from http://www.zin.ru/projects/invasions/ by Vadim Panov, Zoological Institute of the Russian Academy of Sciences, 199034 St. Petersburg, Russia; E-mail: gaas@zin.ru

Chapter 3

PREVENTION

Summary

This chapter provides an extensive, though not exhaustive, list of pathways for alien species introductions and suggests methods for intervention. The majority of exclusion methods were developed for economic pests in the agricultural and forestry sector and would need adaptation to include environmentally important alien species (see Figure 3.1). Finally, potential for successful implementation and the drawbacks of risk assessment procedures are discussed.

The pathway section is classified into four major groups:

➤ Species that are introduced deliberately for use as crops, ornamentals, and game species. A high percentage of vertebrates and plants have been introduced intentionally.

➤ We consider species introduced into captivity separately, although they could be combined with the first group. Many vertebrates become naturalized after escaping into the environment.

➤ Accidental introductions are a major pathway for invertebrates of terrestrial, freshwater, and marine environments. The most significant vector for marine bioinvasions is the ship, either inside the ballast water tank or outside, attached to the hull.

➤ Vectors for spread after an initial introduction are discussed in a final section. It focuses on human structures and alterations of habitat that enhance or enable species to spread within a country or to neighbouring nations.

The experience and expertise of the agricultural and forestry sectors in exclusion methods need to be used as a knowledge base to adapt measures for invasive species in general. Three major exclusion measures to stop introductions are recognized: interception, treatment and prohibition. The first involves the successful implementation of regulations at the border. A risk assessment should be carried out for every proposed intentional introduction. Species whose entry is either Permitted or Prohibited need to be listed in a pied list to allow dissemination of the results of such assessments. Next, commodities suspected of being contaminated with non-indigenous organisms need to be treated, and some treatments are discussed briefly. Finally there is the possibility of prohibiting imports based on international regulations. Education is a key component of all prevention efforts.

The final section discusses the risk assessment process as a tool to support exclusion of species based on their perceived risk and to assess the potential impact of species already established. The objective of such an assessment is to predict whether or not a species is likely to become established and be invasive and to generate a relative ranking of risk. Entire pathways may also be analysed for risk, and this may be a more efficient procedure where many possible species and vectors are involved. The major drawbacks are the uncertainties involved in predicting species' behaviour under different circumstances in a new environment. Thus, lack of knowledge and ability to predict consequences may lead to substantial reliance on assumptions. On the other hand, risk assessment provides a logical process for gathering, analysing, synthesising, comparing and communicating information, which can improve the quality of decision-making.

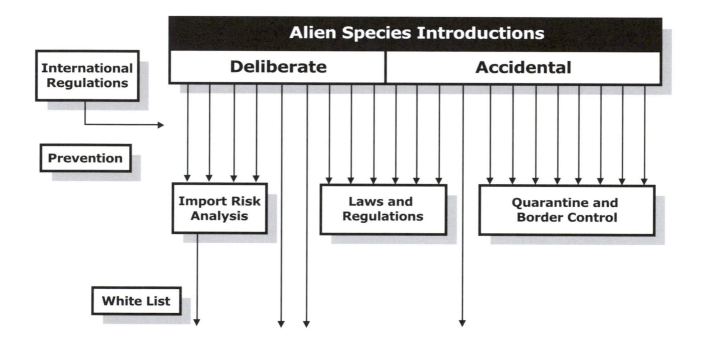

Figure 3.1 - Options to effectively deal with introductions of alien species. Three groups of species will pass through this prevention scheme into the country (depicted as arrows reaching the bottom of the chart). Species accepted on the white list and are authorised for introduction, others slip through the border control, and some are either smuggled directly or enter as contaminants of smuggled items (see figure in the Toolkit Summary for the full flowchart).

3.1 Introduction

An ounce of prevention is worth a pound of cure – this maxim of medicine, dictating such measures as quarantine and inoculation, is equally valid for biological invasions. Prevention is the first and most cost-effective line of defence against invasive alien species.

There are significant costs associated with prevention. One cost, and the most obvious, is the expense of maintaining the exclusion apparatus (salary and training of interception personnel, plus facilities such as fumigation chambers, inspection apparatus, and quarantine quarters). A second cost is that borne by individuals who wish to profit by bringing in alien species (which may or may not be intended for release to the environment). A third potential cost may be borne by a public who have to withstand a further regulatory control or who might have benefited from a planned introduction disallowed by the prevention apparatus.

Against these costs must be tallied benefits that accrue to society from invasions that are prevented. These benefits of an event that did not occur can be difficult to evaluate and even more difficult to portray. But such portrayal and reinforcement of the benefits of not having a particular invasive species are a very important part of publicity surrounding invasive species prevention.

With respect to planned introductions, it is important to emphasize that many of the invasions that have caused the most economic and environmental damage were intentional and planned. Thus the predatory Central American snail, *Euglandina rosea*, liberated on many Pacific islands to control the agricultural pest *Achatina fulica*, has caused the extinction of at least thirty endemic island land snail species and subspecies (Case Study 3.1 "Rosy Wolfsnail, *Euglandina rosea*, Exterminates Endemic Island Snails").

Many plant species, e.g. trees, such as eucalypts introduced for timber production, or other new resources become invasive in natural habitats. Other species that were deliberately imported but not planned for release to the environment have contrived to escape and caused monumental damage. The European gypsy moth (*Lymantria dispar*) escaped from an experimental rearing programme intended to produce a better silkworm and has ravaged forests of the north-eastern United States for a century, while Africanised honeybees escaped from scientific rearing facilities in Brazil and have invaded Central America, Mexico, and the United States, causing deaths and greatly complicating beekeeping (Case Study 3.2 "How Africanised Honey Bees Came to the Americas"). The fact that an introduction is planned, even by scientists, therefore does not mean that it will necessarily be beneficial. Had *Euglandina rosea, Achatina fulica*, the gypsy moth, or Africanised honeybees been intercepted at the outset, an enormous loss of species and/or money would have been prevented.

When considering any deliberate introduction, it should be assumed that, unless there is very clear evidence to the contrary, the introduced species would escape into the wild. Hence, if the species has the potential to become naturalized and invasive, then it will do so.

Of course, with respect to a planned introduction prevented at the outset, it may turn out that, had it been permitted, it would not have become invasive. That is, our ability to predict which species will become invasive, and what impacts they will have, is imperfect (Case Study 3.23 "Invasiveness Cannot Be Reliably Predicted"). However, if a species is permitted and performs as planned (e.g., a new crop, ornamental plant, or pet species) often the benefit largely accrues to the importer, while the cost if it becomes invasive is borne by society as a whole. The national interest would thus dictate a policy of prevention under the precautionary principle (Case Study 3.3 "The Precautionary Principle").

For unplanned introductions, the benefit/cost ratio for exclusion is high. If only a small fraction of inadvertently introduced species become established and only a small fraction of these become invasive pests, this is still an enormous cost to society, far outweighing any unexpected benefits that unplanned introductions might confer and the cost of maintaining the exclusion apparatus. Had the zebra mussel (Case Study 3.4 "The Impact of Zebra Mussel on Ecosystems") been prevented from entering North America by purging of ballast water on the high seas, or by treatment of ballast water by chemicals or ultraviolet light, billions of dollars of industrial damage (from clogged water pipes and other consequences of incrustation) would have been avoided, and the threat of extinction to many aquatic species (primarily invertebrates) would be far lower than it now is. The on-going cost of purging ballast water at sea, or of chemical or ultraviolet light treatment (see "Treatment technologies" in section 3.3), is not negligible, but it pales in comparison to the ongoing monetary, commercial, public nuisance and natural ecosystem conservation cost.

The Asian longhorned beetle is currently established in New York and Illinois, having arrived in wooden packing material from China (Case Study 3.5 "Asian Longhorned Beetle, a Threat to North American forests"). It is a threat to vast tracts of hardwood forests in much of the United States and also to innumerable shade and fruit trees in yards and lining streets. Had it been intercepted at the outset (by inspection in China or North America, by fumigation, or even by exclusion of all wooden packing materials), an on-going cost would have been incurred, but again it would have been far less than the ongoing cost that may now be generated by the beetles' activities.

An important feature of invasions that argues strongly in favour of prevention is that, once an introduced species has become established, particularly if it has become invasive, it becomes extremely difficult to eradicate it. Most attempts to eradicate such species fail, and even failed attempts are often expensive e.g., "the failed 22-year campaign to eradicate the South American fire ant (*Solenopsis invicta*)" from the south-eastern United States cost 200 million dollars (Case Study 5.10 "Fire Ant: an Eradication Programme that Failed"). Once eradication has failed (the usual result), society is faced with damage and management costs in perpetuity. Such costs may be staggering (see e.g. Case Study 3.6 "The Threat of South American Leaf Blight to Rubber in Malaysia"), and this is why stringent prevention measures are warranted.

3.2 Pathways

The most common approach for prevention of invasive organisms is to target individual species. However, a more comprehensive approach is to identify major pathways that lead to harmful invasions and manage the risks associated with these. Although international trade and travel are believed to be the leading cause of harmful unintentional introductions, there is no detailed knowledge base on the actual pathways, except in very few countries. Exclusion methods based on pathways rather than individual species should be a more efficient way to concentrate efforts where pests are most likely to enter national boundaries and avoid wasting resources elsewhere. Moreover, it identifies more species, including more false negatives of the common approach, and identifies more vectors, pathway systems, and underlying introduction mechanisms. Risk assessments can be done for pathways as well as individual species (see Section 3.4).

The argument that some pathways were so extensively used without any prevention methods for decades or even centuries, e.g. ballast water and hull fouling, that all invasive species are already spread to all potential areas is deceptive. Cases, where alien species are introduced for decades but failed to establish until recently, prove that the establishment rate can vary over time. Reasons will include changes in the alien species itself, changes in the pathway (shorter passage time of transatlantic ship traffic increases the likelihood of survival for ballast water species), climatic changes, and changes in human impact in the area of introduction (salinity and nutrient changes in bays etc.). The accelerating rate of establishment of alien species demonstrates that the concerns about accidental introductions are still valid.

This section provides an overview of potential entry pathways for alien species. Thus, the list of pathways and invasive organisms is not meant to be comprehensive. Besides, future pathways will be created with every new invention in mobility and trade. Moreover, an exhaustive list of pathways will be produced by another programme element of GISP, i.e. pathways/vectors of invasives (details at the end of this section).

Most of the knowledge on early introductions (e.g. pre 1950) is in anecdotal form rather than officially recorded (Case Study 3.7 "Anecdotes about Entry Pathways") and many of the more recent introductions are only poorly documented. Detailed reporting of new accidental and deliberate introductions in the local official or scientific literature should be encouraged. This should include the source, method of entry and the fate of the introduction; it should also make clear what is fact, what is deduction and what is speculation.

A list of major pathways with examples and potential prevention methods against invasive organisms is given in the following section. The vectors are summarized under two categories, i.e. intentional and accidental introductions. The intentional introductions are split into two different modes of naturalisation: species directly released into the environment and escapes from containment into the environment. Most plant and vertebrate species introductions have been intentional for various reasons, e.g. plants as ornamentals, mammals as game, birds as delight for the spirit and the senses, fish for sport fishing. On the other hand, most invertebrates (including marine organisms) and microbe introductions have been accidental, often attached to other species introduced intentionally. Often agricultural weeds have been introduced as contaminants of crop seeds, whereas most of the environmental weeds were purposefully planted as ornamentals, for soil stabilization, for firewood, etc. sometimes supported by ill-guided aid programmes or commercial ventures. All 13 declared noxious weed species of French Polynesia were introduced intentionally as ornamentals, or for other purposes.

It should be stressed at this point that education is a key component of successful prevention and management methods. The public has to be informed why prevention measures are taken and what impact failure can cause. The public as well as the companies concerned should perceive prevention measures not as arbitrary nuisance but rather as necessary aspects of travel and trade to care for the future commercial and natural environment.

For more comprehensive information on pathways, please refer to a document prepared by another GISP group that focussed on pathways of invasive alien species: *Gregory M. Ruiz and James T. Carlton, Editors, 2001. Pathways of Invasions: Strategies for Management across Space and Time. Island Press, Washington, D.C. (Volume in preparation)*. This is the symposium volume arising out of the November 1999 GISP conference on pathways.

3.2.1   Intentional introductions

We consider first pathways based on direct introduction of invasive species into the environment.

Plants introduced for agricultural purposes

Foreign plants are introduced for a great variety of purposes. A large proportion of important crops are grown in areas outside their natural distribution for economic reasons, to diversify national agriculture, and as a safer way to feed the world population by spreading the risks of disease outbreaks. If a new alien crop is introduced without its pests, this "pest-free" species can be especially productive and profitable. On the other hand these foreign species can pose a risk to biodiversity when they naturalize and penetrate conservation areas encroached by these fields. Tall fescue (*Festuca arundinacea*), a native grass of Europe has been planted as a pasture grass in North America; it has naturalized and invaded remnant prairies, replacing the diverse natural herbaceous community.

In order to prevent a future invasion by a proposed new crop, an import risk analysis is required (considered in more detail in Section 3.4). Such an analysis assesses the risk of establishment, spread, and impact of the plant under consideration and is prepared by consulting with stakeholders and technical experts. A national panel balancing the risks and the potential advantages would take a final decision on whether to proceed with importation.

Foreign plants grown for forestry use

The situation in the forestry industry, whether promoted by a government, a commercial user or an aid programme, is similar to the agricultural sector. A significant number of forestry tree species, including agro-forestry and multi-purpose tree species, have become invasive as aliens. New rapidly growing and less labour intensive trees are continuously sought to enhance timber production. Foreign trees are planted in plantations, often of huge dimensions. Many of these exotic trees become established and spread into natural habitats, displacing the native vegetation. For example, in South Africa, *Pinus, Acacia* and *Eucalyptus* species are the basis of an important and lucrative industry, but they are also a substantial threat, as invasive alien plants, to major conservation areas and to the country's meagre water supplies.

Species regarded as invasive could be prohibited in a country or the spread from plantations could be minimized by controlling the invasive species around the plantation as they start to spread. Regulations should be developed so that the owner of the plantation, who has the benefits from the business, has to carry out control measures.

Non-indigenous plants used for soil improvements

Exotic plants are frequently planted for improvement of soil characteristics (e.g. plants with nitrogen fixing abilities), or for erosion control and dune stabilization. In the 1930s the U.S. Soil Conservation Service grew millions of seedlings of kudzu (*Pueraria lobata*) and sold them to farmers to grow to prevent erosion. Scotch broom (*Cytisus scoparius*), native to Europe, is another example of a plant promoted and used to prevent erosion and to stabilize dunes in North America. It currently covers more than two million acres of grasslands, scrublands and open canopy forest in the western states. This foreign plant is a threat to humans, livestock, and native plant species in the invaded areas. Scotch broom is flammable and carries fire to the tree canopy layer where fires burn hotter and are more destructive. It displaces the native plant cover, particularly in soil characterized be a deficiency of nitrogen, due to its symbiosis with nitrogen-fixing bacteria in the root nodules.

"Aid-trade"

Aid programmes need to consult conservation authorities to prevent introductions of alien organisms, which might have a short-term benefit, but prove to be a biodiversity threat and inflict long-term costs exceeding the initial benefit (see also as an example of contamination: Case Study 3.10 "The Introduction of Parthenium Weed into Ethiopia").

The Central American tree *Cordia alliodora*, planted as a potential timber plantation tree on Vanuatu is a classic example of a problem arising from an introduction by an aid programme. It was introduced with the best intentions, but failed to live up to expectations for various reasons, probably linked to climatic differences between Central America and Vanuatu. *C. alliodora* became a nuisance, slowly penetrating the native bush. Other well-known examples include promotion of pines and eucalypts around the world as timber trees in new environments.

Ornamental plants

A high percentage of plant invaders were originally introduced as ornamentals. The South American *Lantana camara*, one of the most invasive and widespread of tropical weeds in the Old World, was spread throughout the tropics in a variety of hybrid forms as a garden ornamental. About half of the 300 most invasive plants in North America were introduced to gardens and parks as ornamental plants. Old man's beard (*Clematis vitalba*), a European vine, was planted in gardens and parks in New Zealand in the 1930s. Thirty years later it was recognized as a threat to native remnant forests, where it smothers even mature trees (as an example see also Case Study 3.11 "Long-distance Spread of *Miconia calvescens* to Remote Islands of French Polynesia").

It can be argued that such introductions as well as trade in invasive plants or species related to weeds should not be permitted, in which case importers would need to prove that species are environmentally acceptable before they could be imported. The plant growing industry has developed a strategy to sell invasive plants as non-fertile specimens under the pressure of biodiversity awareness.

Germplasm

Germplasm importation for propagation was a recognized pathway by which several plant pathogens were introduced into new areas. Increasing availability of tissue culture should prevent such problems in the future.

Birds and mammals released for hunting purposes

In historical times mammal and bird species were widely introduced by new settlers to maintain a hunting tradition with familiar game animals. Several deer species have been transferred to new locations all over the world. Deer alter habitats by preferential browsing on selected plant species. Many plant species on islands without large native mammals are not adapted to damage caused by such ungulates. New introductions of species for hunting purposes should at least involve a risk assessment process (see Section 3.4). Invasive species already present can sometimes be controlled by effective hunting and can be eradicated on small islands by shooting.

Mammals released on islands as food resource

During the times of sailing ships roaming the oceans, it was a common habit to release farm animals, such as goats, pigs, etc., on uninhabited islands as a food resource for subsequent visits or for the benefit of shipwrecked sailors (Case Studies 3.7 "Anecdotes about Entry Pathways" and 5.7 "Rabbit Eradication on Phillip Island"). These animals are still thriving without natural enemies on many of these islands. Their browsing pressure, especially at high populations puts a significant number of plant species at risk.

Biological control

Introductions of organisms for biological control, particularly several cases in earlier projects, have on occasion caused damage to non-target species. Most of these examples involved the introduction of generalist predators, often vertebrates (see section 5.4.3). The small Indian mongoose was released on many islands, including the Hawaiian Islands, from the end of the 19th century through to the 20th. Instead of controlling introduced rats, they found easier prey and devastated the islands' native bird fauna, especially the ground-nesting birds. Similarly, the cane toad, introduced in an attempt to control beetle pests in sugar cane in Australia, preferred native amphibians and a wide range of invertebrates as food. The toads also poisoned pets when they came in contact with the toads' poisonous skin. Soon after release the cane toad reached high abundance levels and has since been considered a problem itself (Case Study 5.39 "A Preliminary Risk Assessment of Cane Toads in Kakadu National Park").

Another example of unwanted non-target effects resulting from a biological control introduction involves the predatory snail *Euglandina rosea*, introduced into many Pacific islands to control the giant African snail, which was itself a misguided introduction as a food resource, and became an agricultural pest (Case Study 3.1 "Rosy Wolfsnail, *Euglandina rosea*, Exterminates Endemic Island Snails").

Today the safety standards for biological control are far more rigorous and are regulated by laws and regulations, such as the International Plant Protection Council's "Code of Conduct for the Import and Release of Exotic Biological Control Agents". All biological control projects should have a scientific basis and a risk analysis (see Section 3.4) conducted before an application for import can be submitted. In most cases the need to use highly host specific agents, would exclude vertebrates from use. All stakeholders need to be involved in the decision-making process. Biological control agents are also normally quarantined upon importation, to screen for contaminants such as parasites and diseases, and to check the purity of the material. Nevertheless, it should always be taken into consideration that any introduction is a permanent decision, and that a successful biological control agent will spread, perhaps to unanticipated areas (Case Study 3.8 "Spread of a Biological Control Agent, *Cactoblastis cactorum*, in the Caribbean Basin").

Fishery releases

Beside releases into containment of aquaculture, fish species are often released into the wild to expand the recreational fishery (Case Study 3.9 "Release of Exotic Fish by Aquarium Hobbyists – the USA Experience"). The European brown trout has frequently been introduced as a game fish in North America and in highland rivers and streams through much of the tropics. It is suspected of contributing to declining numbers of native fish species through direct competition. Regulations are needed to control releases in order to prevent additional invasive fish species.

Pets released into the wild and aquarium trade

Pets and aquarium inhabitants, if not wanted any more for any reason, are often released "back" into the wild with well-meant intention (Case Study 3.9 "Release of Exotic Fish by Aquarium Hobbyists – the USA Experience"). Terrapins, crocodiles, aquarium fish and flora released into ponds and down toilets, easily find their way into the local water system. Public education is the major tool to minimize these releases. The owner has to be informed that the species are exotic and either cannot survive in the new environment or will survive and pose a risk to native species. In addition, the trading organizations should be required to accept the organisms back. In the case of pet species not already in a country then stringent import rules based on risk assessment (see Section 3.4) should be applied as for other introductions.

Reintroductions

Under this heading, cases of introductions of species that are closely related can be included, since definitions of species and sub-species are often somewhat blurred. One example of an attempt to reintroduce a rare mammal back into its natural habitat in Europe was the release of beavers. However, the released animals were of North American origin, where they were fairly common, and now are recognized as a different species from the European beaver. Another example is the introduction of American crayfish after the closely related European crayfish population crashed due to a highly specific disease outbreak. Only specimens of known origin from the same or nearby populations should be considered for reintroductions into the wild. Any taxonomic difficulties have to be resolved beforehand. Subspecies from a different part of the range can be considered for release if the native subspecies is extinct. However, if the native subspecies is rare but a population is still surviving, the specific genetic material adapted to that location would probably go extinct through breeding with the new releases.

Releases to "enrich" the native flora and fauna

Many intentional introductions were purely sentimental. When people settled in new territories they tried to make the new environment more familiar and comfortable by releasing popular and attractive species, such as flowers and birds, from their home country. Starlings and house sparrows, amongst other species, have been introduced from Europe into many countries, leading to losses of native songbird populations through direct competition for food or nest space. Legislation and international regulations regarding exotic species are needed in many countries or, where they are in place, they need to be implemented or need enforcement to counter illegal introductions. The large number of such useless intentional introductions strongly supports a shift in policy from the more conservative approach of blacklists to a more stringent policy of "guilty until proven innocent". People acquire an aesthetic appreciation for introduced beautiful flowers and birds, and often become opposed to eradication programmes against these species (Case Study 3.21 "Two Views of the Rainbow Lorikeet in New Zealand").

3.2.2   Introductions to captivity

The following pathways are examples of routes by which species have been intentionally introduced only to a containment situation but which subsequently escaped into the environment.

Escapes from captivity such as zoos and botanical gardens

Alien species introduced into containment in a new country may escape from captivity and become invasive. Completely effective fences to contain mammals are prohibitively expensive and often a cheaper but less effective solution is chosen. Even fences regarded as 100% secure, are not immune to accidental or deliberate damage by humans, e.g. animal rights groups or damage by natural events such as trees falling on the fence or tornadoes destroying the enclosure. In many parts of the world, including treasured island habitats, wild boars of European origin either intentionally introduced for hunting or escaped from hunting enclosures are altering the entire character of ecosystems. They change the composition of local plant communities by feeding selectively on plants with starchy bulbs, tubers, and rhizomes. Additionally, the pigs have a tremendous impact on the nutrient flow between the soil layers by their digging activities. These disturbances of the plant cover often favour alien plants and enhance the seed recruitment of invasive weeds. Several major alien weed species cannot penetrate undisturbed native vegetation and need disturbances to colonize new areas. Thus, introduced wild boars are important mechanisms for invasion by alien plant species. Reproductive sterilization is perhaps the most secure approach for the biological containment of non-indigenous species, where this approach is an option. Completely effective measures to contain plants in a growing facility can also fail as wind borne seeds may drift out or seeds lodge in clothing to be carried to a new site. Plants can also be so attractive that a staff member will just "take a little bit" or pieces or seeds may wash down a drain to grow in another place.

Farmed mammals

People often consider a new animal to be a potential farming bonanza. Governments like the idea of people making money and paying taxes and so often allow new farm animals to be imported and housed or farmed with minimal restrictions. The assumption is made that it is in the farmer's best interests not to allow the animals to escape. Poor care of enclosures, natural disasters or financial failure often results in the farmed animal escaping to the wild. For example there were no wild deer in New Zealand, but farming behind specially designed high deer fences was permitted in the North. The deer escaped and there is now a need to eradicate them from the valuable native forest areas.

There are several similar examples of escapes in the fur industry. Mink (*Mustela* spp.) are valued for their dense winter fur. When populations of the European mink, *M. lutreola*, were in decline due to fur hunting and habitat loss, American mink, *M. vison*, was introduced into fur farms in Europe in the 1920s. Some American mink subsequently escaped and others were deliberately released into the wild to establish free-living populations to harvest. Also, in recent years, there have been several instances when animal rights activists have raided fur farms and freed the inhabitants. American mink was soon established in several places of

Europe and has increased rapidly in numbers. It is a predator of fish, birds, mammals and smaller food items. Its activities, together with habitat destruction, have brought the native water vole to the brink of extinction in the UK. The larger American mink replaces its European relative by competition and interbreeding. In spring it is sexual active earlier, so that the male American mink mate with European females. They do not produce fertile offspring but the European females thus mated are effectively excluded from breeding (Case Study 3.25 "Eradication Programmes against the American Mink in Europe").

Aquaculture and mariculture

Non-indigenous species are frequently used in aquaculture and mariculture. Escapes from marine net pens are not uncommon and escapees often invade the new habitats (see also Case Study 3.9 "Release of Exotic Fish by Aquarium Hobbyists – the USA Experience"). Approximately 80% of the salmon production on the Canadian Pacific coast is based on an alien species, the Atlantic salmon (*Salmo* salar). The continuous addition of adult Atlantic salmon into the coastal environment may affect the population of the native relatives, Pacific salmon (*Onchorhynchus* spp.); recent observations suggest a successful colonization by the exotic species. Since containment in aquaculture cannot be guaranteed, species should not be introduced until a risk assessment (see Section 3.4) has been undertaken to assess the safety of the action proposed.

Research and introductions through research institutes

This is not a major pathway, but there are some very significant examples. The Africanised honeybee escaped from a research facility in Brazil and spread through the Americas (Case Study 3.2 "How Africanised Honey Bees came to the Americas"). Another case of a destructive escapee is the gypsy moth, which was being held in containment in the hope of breeding a new species for silk production. Where research on non-indigenous species has to be carried out in a country, it needs to be licensed, and alien species kept under strict containment measures. An alternative option, which should be preferred for the study of high-risk species, is that the researcher travel to work in the natural range of the species to be studied.

3.2.3 Accidental introductions

Contaminants of agricultural produce

Fruits and vegetables can harbour a wide variety of immature stages of insects, most notably fruit flies in a variety of fruit species. Treatment techniques for known pest species are routinely applied. However, not all imported produce is treated, and invertebrate species, in particular, frequently reach ports via this pathway, as newspaper reports of bird-eating spiders emerging from banana boxes convincingly demonstrate.

Emergency aid can result in food and materials being rapidly moved around the world, often straight into rural areas. Famine relief activities have on several occasions been implicated in the introduction of non-indigenous species as contaminants of food grain (Case Study 3.10 "The Introduction of Parthenium Weed into Ethiopia").

Seed and invertebrate contamination of nursery plants

Besides the threat to biodiversity posed by intentionally introduced plant species themselves, imported plants can be contaminated with other organisms. Species living on or in imported plants are another major source of invertebrate introductions. The Stenorhyncha bugs comprising mostly sedentary groups such as scale insects and mealy bugs are particularly prone to be dispersed in this way (Case Study 5.11 "Colonization Rate of Hibiscus Mealybug in the Caribbean"). Seeds of other plant species can be attached to the plant material.

Seed and invertebrate contamination of cut flower trade

The transfer of invertebrates on live plants applies as well to the cut flower trade. Leaf-miners (e.g. Agromyzidae), thrips, mites and larvae of several moth species are regularly found on cut flowers indicating the risks associated with the inter-continental flower trade and the importance of treatment methods to minimize risks. Quite apart from the trade through normal channels, often one can observe passengers on aeroplanes carrying bunches of flowers, perhaps picked from a local garden with little or no insect pest control only hours before they are carried onto the plane and taken to another country. This is one of the mechanisms implicated in the spread of hibiscus mealybug (Case Study 5.11 "Colonization Rate of Hibiscus Mealybug in the Caribbean") and there are many species that could be moved in this way.

Organisms in or on timber

Timber is also a breeding substrate for a huge number of invertebrate species, including many beetle species. Unprocessed wood and wood products are a source of forest pests and pathogens. Strict regulations on importation and measures to clean the material are necessary (Case Studies 3.5 "Asian Longhorned Beetle, a Threat to North American Forests" and 3.22 "Siberian Timber Imports: Analysis of A Potentially High-Risk Pathway").

Seed contaminants

Many of the alien agricultural weed species have been accidentally introduced as contaminants of crop seeds. Despite the Federal Seed Act, weeds continue to arrive in the USA as seed contaminants. It is believed that serrated tussock grass (*Nasella trichotoma*) was introduced from South America into Australasia, Europe and North America in this way. It is capable of replacing native grasslands once they have been disturbed by other means. Improvement of threshing and harvesting machines has reduced the number of seed contaminants. However, seeds of some species were successfully selected in evolutionary terms to resemble crop seeds and are exceedingly difficult to separate. Thus, weed seeds are widely distributed and then planted in favourable conditions along with the desired agricultural seeds.

Soil inhabiting species

Soil-inhabiting species can be introduced by shipping soil or by soil attached to plant material. A formerly significant source of exotic plant and insect species has ceased to be a pathway at the beginning of the century, when ships switched from taking dry ballast soil to ballast water. However, many present-day pests were brought into their new environment via this route. Before the days of airfreight, crop plants were regularly transported on ships as plants growing in soil. Soil pests were undoubtedly spread in this way, although this was often undocumented. *Clemora smithi*, a sugar cane white grub, the larva of a beetle, was transferred from Barbados to Mauritius in just this way. Many living plants are still moved as potted specimens. Without doubt many soil-inhabiting microorganisms are spread around the world using this transport vector.

Machinery, equipment, vehicles, military, etc.

Machinery and vehicles are often shipped from place to place without cleaning. Depending on the nature of their use, they may carry soil and plant material (Case Studies 3.10 "The Introduction of Parthenium Weed into Ethiopia" and 3.11 "Long-distance Spread of *Miconia calvescens* to Remote Islands of French Polynesia").

Historically, military equipment has resulted in several introductions of harmful species, such as the golden nematode (*Globodera rostrochinensis*) into the USA. If military action does not permit cleaning of the vehicles before shipping, the material should be cleaned upon arrival at specially designated places and all material found should be destroyed (Case Study 3.12 "The Australian Defence Force is Involved in Keeping Alien Species Out").

Hitchhikers in or on package material

Stowaways of all kinds have been found on diverse packing materials. Investigations on packing material of bait worms from Asia to North America revealed an active pathway for invasion of many different organisms and possible pathogens. Upon arrival live species of several taxa were found to be on and in the packaging material, including the bacteria *Vibrio cholerae*, the causative organism for cholera (Case Study 3.13 "Hitchhikers Moved with Marine Baitworms and Their Packing Material"). Viable algae and spartina grass have been found used as packing material for transport of oysters.

Another example is wooden material and dunnage, which is believed to be the vector for some exotic bark beetle species. The US Government required China to apply phytosanitary measures to all unprocessed wood packing material, after the second interception of the Asian longhorned beetle in trees near a US port (Case Study 3.5 "Asian Longhorned Beetle, a Threat to North American forests"). This is only one potentially devastating species arriving on wood packing material - insects from 54 families have been intercepted in this material by USDA.

Hitchhikers in or on mail and cargo

Small species such as insects can easily hide in all sorts of cargo. The long-legged ant, *Anoplolepis gracilipes*, has spread throughout the tropics travelling as stowaways in cargo. This ant species forms super-colonies with multiple queens and little territoriality. These unusual features allow high densities and population explosions. The long-legged ants cause havoc through direct predation on invertebrates and even vertebrates, many times bigger than themselves. Only strict quarantine procedures and inspections can prevent introduction of small species travelling as stowaways.

For cargo suspected of contamination with alien species a whole array of treatments is available. The goods and their package material and containers can be treated with pesticides through fumigation and immersion. Other methods include heat or cold treatment and irradiation. Cleaning of the goods and the packages is highly labour and cost intensive, but are essential to prevent introductions. Special reception areas can be useful (see e.g. Case Study 5.34 "Ecotourism as a Source of Funding to Control Invasive Species").

The use of shipping containers offers considerable scope for stowaways, and they are difficult to inspect adequately. In one extreme case a racoon survived for about five weeks in a container while it was shipped from the USA to Europe and was still able to walk out of the container. Containers used to transport raw timber frequently carry many associated species. Even apparently "clean" cargoes can carry invaders such as the scorpions recently transported from Portugal to New

Zealand in new empty wine bottles despite recorded fumigation of the container before departure.

The Asian tiger mosquito (*Aedes albopictus*) was accidentally introduced to the USA from Japan in the mid 1980s; it was transported in water collected in used tires, in which they often breed. This mosquito species attacks many hosts and vectors diseases between wildlife and humans. Introductions of diseases and vectors have mutual effects and in some cases outbreaks of certain pathogens were observed only after the introduction of a suitable vector (Case Study 4.9 "Spread of the Aphid Vector of Citrus Tristeza Virus").

Hitchhikers in or on planes

Some exotic species are able to hitchhike on the outside of a plane, but travel within planes is much more common (Case Study 3.14: Spread of the Brown Tree Snake in the Pacific Region"). Quarantine measures on arrival are difficult to enforce. In general, a more promising approach would be to make sure that the planes do not have hitchhikers onboard before take off (Case Study 3.18: "Sorry, No Free Rides from the Torres Strait"). However, most investment on prevention is for checking imports rather than exports (Case Study 2.15 "Mauritius and La Réunion Co-operate to Prevent a Sugar Cane Pest Spreading" for an exception to this).

Ballast soil

The North American cord grass is an example of a plant believed to have been introduced to Europe as seeds in soil used as ship ballast (Case Study 5.4 "Hybridisation"). Modern ships use water for ballast instead of dry material (see next pathway), so this pathway is mainly of historical significance.

Ballast water of ships

The most significant pathways for marine bioinvasions are in the ballast tanks of ships and the fouling on the outside of ships' hulls. Despite the difficulty of proving that an invasive species has been introduced through a particular pathway, examination of ballast water has demonstrated the enormous potential importance of that pathway. Literally hundreds of species can be found alive in samples from a single ship. It was estimated that on average one tanker releases about 240 million organisms into the surrounding water on each voyage. When ships dump this load of diverse organisms in waters similar to their origin, there is no doubt that species will become established in this new environment. Probably the most infamous introduction via ballast water is that of the zebra mussel into the Great Lakes in North America (Case Study 3.4 "The Impact of Zebra Mussel on Ecosystems").

Moreover, ballast water may pose a substantial threat to human health. Ship-mediated dispersal of pathogens may play an important role in the emergence and epidemiology of some waterborne diseases, such as the bacteria *Vibrio cholerae*, the agent of human cholera. Methods for ballast water treatment are currently under investigation. One measure already in place to prevent further bioinvasions into the Great Lakes is the change from voluntary to mandatory ballast water exchange. In the marine environment this exchange is still on a voluntary basis. Other methods being explored are filter systems at water intakes, irradiation using Ultra Violet Light (UVL), drinking water treatment methods, heat treatment with waste heat of the engines, and dumping ballast water into land-based plants as used for sewage treatment.

Ballast sediment in ballast water tanks

On the bottom of ballast water tanks sediments become concentrated, allowing organisms adapted to these conditions to survive and be moved from place to place. Whereas the ballast water favours pelagic species, the sediment hosts ground-dwelling species and increases the number of species able to survive the journey between the intake and the dumping of the ballast water. Ballast water can be exchanged during the voyage, but the sediment is not flushed out. Thus, more rigorous methods are needed to treat ballast sediment. Besides using chemical and heat treatment, ballast water tanks should be cleaned on a more regular and frequent basis.

Hull fouling

Fouling organisms on ship hulls have caused economic losses since the first ships sailed the oceans. The greatest invasive species risks are associated with ships and machinery kept in ports for some time and then transferred to a new destination. Several cases came to light when docks were moved to another port and upon arrival several hundred species were found living on the hull.

The ranking of importance of the last three mentioned pathways for marine organisms are vehemently debated and the specific vector for most marine bioinvasions is to some extent a matter of guesswork. However, there is little doubt that movement of ships is the most import pathway in the movement of marine organisms from country to country and from sea to sea (Case Study 3.15: Monitoring for the Black Striped Mussel in Northern Territory, Australia"). The costs associated with cleaning procedures for ship hulls seem to be indispensable.

Debris

For a long time, marine debris has been known to pose environmental threats due to wildlife entanglement and ingestion. It is also considered an aesthetic factor that can influence tourism. In addition pelagic plastic can function in a similar way to ships' hulls to transport organisms. Studies have revealed that persistent floating synthetic materials often support a varied community of encrusting and fouling epibionts as well as attracting a diverse motile biota. Thus, pelagic plastic plays a role as surrogates for the substrata provided in nature such as floating seaweeds, logs, and free-swimming marine animals. A survey of marine debris in northern New Zealand waters revealed 28 of 60 bryozoan species that had not previously been recorded. Reducing the prevalence of synthetic material in the oceans would require a change in the attitudes of the public and industries involved.

Tourists and their luggage/equipment

The dramatic increase in tourist volume and mobility is swiftly increasing in importance as a vector for introduction of alien species into remote areas. The trend for new outdoor activities and sports is leading to more rapid movement of tourists and their equipment into the remotest corners of the globe. Public awareness of the problems involved with bioinvasions and public education about how to behave are considered an essential element of prevention programmes. Education before travellers depart offers perhaps the best way to prevent introductions, by allowing them to clean their equipment and leave prohibited items behind. Broadcasting of educational films on the plane is a good way to raise the awareness of the invasives problem (Case Study 3.18 "Sorry, No Free Rides from the Torres Strait"). Laws to prohibit exportation and importation of organisms (as souvenirs etc.) need to be in place and enforced. People not only transport species on soil-contaminated equipment etc. accidentally, but many tourists bring home plants, plant parts or live animals as souvenirs.

Tour operators also need to be involved. It is in their own self-interest not to allow the habitats to which they take tourists to be spoilt by invasive species. Moreover, tour operators should be required to take responsibility for the behaviour of their tourists.

Diseases in animals traded for agricultural and other purposes

The disease brucellosis was probably introduced into the USA in imported cattle and now causes major economic losses in domestic livestock as well as infected bison and elk. Distemper from domestic dogs has been linked with outbreaks of the disease in populations of the endangered African hunting dog. Sanitary measures and inspections have to be in place and enforced to decrease the risks by importation of animal diseases.

Parasites, pathogens and hitchhikers of aquaculture and mariculture

Any movement of species used in aquaculture and mariculture carries the risk of transferring parasites and diseases (Case Studies 3.16 "Transfer of Pathogens and Other Species via Oyster Culture" and 3.17 "Japanese Brown Alga Introduced with Oysters").

Even trade in indigenous animals can lead to accidental introductions of pathogens, when the trade movement includes foreign places where they can pick up diseases. When the North American rainbow trout was introduced to Europe, it suffered epidemics of whirling disease, caused by an indigenous parasite, which had broadened its host spectrum from the European brown trout to the new arrival. Subsequent indiscriminate transport of rainbow trout between breeding facilities spread the disease to other parts of the world, including North America. Sanitary inspections of shipments before or after arrival can minimize the risks.

3.2.4   Vectors of spread after introduction

This sections focuses on mechanisms and circumstances that enhance spread after introduction of a species into a new environment has occurred. Quite apart from initial intentional releases or accidental introductions, many species then spread within a country or cross-national boundaries. Some species undergo an explosive expansion of their range after "barriers" are removed or new pathways are opened by human activity even if the initial introduction has happened a long time ago. Knowing these natural barriers are the basic knowledge used in containment programmes of introduced species (see Section 5.3.2).

Spread from neighbouring countries after introduction

An invasive species after introduction to a new environment will spread into neighbouring countries where there is suitable habitat. This raises the questions of responsibility and liability. International regulations ratified by the neighbouring countries can reduce the risks of bioinvasions by agreed measures. Regional initiatives will frequently be needed to exclude invasive alien species, and to manage them once they are established.

Human-made structures which enhance spread of alien species

Structures which link otherwise unconnected freshwater bodies, marine bodies, or landmasses are an important pathway – not only for invasive alien species but also for indigenous species to reach new watersheds. The completion of the Welland Canal between Lake Ontario and Lake Erie, enabled invasive organisms, such as the sea lamprey (*Petromyzon marinus*), to bypass the Niagara Falls and subsequently spread to other lakes and river systems. The opening of the Suez

Canal initiated a remarkable influx of hundreds of Red Sea species into the oligotrophic Mediterranean Sea, outcompeting and replacing indigenous species. Electric barriers to stop the spread of invasive species along canals are amongst approaches currently under investigation. Successful measures to stop the spread of invasive species along canals seem to be difficult to implement, but if successful could be incorporated at the construction of new canals.

Human alteration of habitats and changes in agricultural practices

Most pest species become invasive after a considerable lag time during which they persist in small numbers until there are outbreaks and invasion starts to occur. Several causes for this delay are discussed in the literature. One is the change of conditions in the ecosystems caused by human land use changes or changes in agricultural practises, which may favour some species over others, or constructions of new pathways linking habitats, etc. Suddenly these species can increase in population size and can become invasive.

3.3 Exclusion methods

Most of the current prevention measures target certain species known to be pests in the country or elsewhere. However, these species are predominantly economically important species for the agricultural, forestry, or human health sectors. Prevention of species on these "black lists" is the rather conservative goal for quarantine and other measures taken at present. A more recent approach in order to incorporate all potentially dangerous organisms, not only in an economic view but also in terms of saving the world's biodiversity, is a move to using "white lists". The approach is often also called "guilty until proven innocent". A proposed intermediate step is the use of "pied lists", favoured for reasons described below.

Since present facilities and staffing are inadequate to process the high volume of incoming material in respect to all organisms accidentally introduced with it, different species or taxonomic groups of potential invasive species have to be treated in different ways, since treatment technologies are often species-specific. Thus new technologies addressing all organisms using a specific pathway are not yet available and need to be designed.

The most reliable method for predicting a species' invasiveness, is to extrapolate from its record as an invasive species under similar conditions elsewhere (see also Section 3.4 and Case Study 3.23 "Invasiveness Cannot Be Reliably Predicted"). Species known to be invasive elsewhere must be considered high priority black list species, like the brown tree snake is for Hawaii. The "pied list" would contain a section of known pest species (equivalent to black lists) with strict regulations and measures to ensure pest-free imports. Another section of the list would describe species cleared for introduction (white lists) – organisms declared as safe. All species not listed are regarded as potential threats to biodiversity, ecosystems, or economy. A stakeholder proposing an intentional introduction has to prove the safety of the species in a risk assessment process before introduction (cf. Section 3.4). Species assessed for their likely invasiveness would be added to the white or black list depending on the outcome of the investigation. However, since invasiveness of alien species can vary with time, genetic composition of the introduced population, and changes in human behaviour (e.g. in land use), the species on the white lists have to be re-assessed in appropriate intervals, e.g. environmentally benign species can become invasives.

There are three major possibilities to stop further invasions.

1. **Interception.** The first step is based on regulations and their enforcement with inspections and fees. **Accidental introductions** are best addressed before exportation or upon arrival of goods and trade. This approach involves decontamination, inspection or constraints to specific trade rated as high-risk commodities. Another approach has to address illegal import of prohibited items - **smuggling**. Whenever regulatory laws are set in place, it seems some people will try to evade them. There are of course no inspections of smuggled items and smugglers are unlikely to sterilize their goods, so this pathway is definitely high-risk for introductions. Staffing and financial constraints set limitations to the prevention of smuggling. In order to meet the precautionary principle a risk assessment process should be the basis for every proposed **intentional introduction** unless the species is listed in the white part of the pied list (cf. Section 3.4).

2. **Treatment.** If goods and their packaging material are suspected to be contaminated with non-indigenous organisms or where high security is required for other reasons, treatment is necessary. That may involve biocide applications (e.g. fumigation, pesticide application), water immersion, heat and cold treatment, pressure or irradiation.

3. **Prohibition.** Finally, when even strict measures will not prevent introductions through high-risk pathways, trade prohibition based on international regulations can be set in place. This can be applied with respect to particular products, source regions, or routes. Under the World Trade Organization – Sanitary and Phytosanitary Agreement (WTO SPS Agreement) member countries have the right to take sanitary and phytosanitary measures to the extent necessary to protect human, animal, or plant life or health provided these measures are based on scientific principles and are not maintained without sufficient scientific evidence.

In the following sections, technologies and methods for prevention of entry and establishment of non-indigenous species are outlined.

Quarantine laws and regulations

International trade requires the establishment of regulatory quarantine. Laws need to be established and enforced. Due to staff and financial limitations quarantine laws are often not adequately enforced and are then useless as a prevention tool. In many cases there is a need to incorporate environmental pests into these laws, in addition to their current major focus on agricultural pests. There are several international regulations focusing on invasive organisms and global trade, including the WTO SPS agreement, the International Plant Protection Convention (IPPC), and the Office International des Epizooties (OIE). The WTO SPS Agreement defines the basic rights and obligations of WTO member countries with regard to the use of sanitary and phytosanitary measures, which are measures necessary to protect human, animal or plant life or health, including procedures to test, diagnose, isolate, control or eradicate diseases and pests. The IPPC develops international standards for phytosanitary measures, e.g. "Code of Conduct for the Import and Release of Exotic Biological Control Agents". The OIE is establishing animal health standards and guidelines for international trade in animals and animal products.

Accessibility of information on invasive organisms

Customs controls can establish useful databases on non-indigenous species encountered at their borders including information on what species are found, by which route they arrived, and what pathway was involved. If these databases are available for other countries, prevention measures could be made more effective on a global scale. However, responsibility for the information listed in the databases can restrict their practicality. It is possible that countries will be reluctant to admit the occurrence of specific pests within their borders, fearing subsequent trade restrictions. In addition, databases of species known for their invasiveness with information on distribution, pathways, management options, etc. are available from many organizations. GISP has developed a database related to this toolkit available at http://www.issg.org/database (Case Study 3.24 "GISP Global Database / Early Warning Component"; see also Box 2.1 "Some Internet-based Databases and Documents on Invasive Alien species").

Public education

Public education is an essential part of prevention and management programmes. In fact, some scientifically well-devised projects have been interrupted or stopped because of lack of public approval. Besides these extreme cases, public awareness and support can increase greatly the success of projects to protect and save biodiversity. Travellers are often unaware of laws and regulations to prevent introductions of alien species, and the reasons for them. Education should focus on raising the awareness of the reasons for the restrictions, regulatory actions, and

the environmental and economic risks involved. In addition to printed material, e.g. posters and brochures, video presentations and announcements on airplanes are a promising approach (Case Study 3.18 "Sorry, No Free Rides from the Torres Strait"). This approach can give the traveller the opportunity to react upon arrival, e.g. using honour bins for prohibited items.

Inspection

Prevention is considered the most economical, desirable, and effective management strategy against harmful invaders. The manifestation of this policy is border inspection and exclusion programmes. Introduced plants and animals are checked for diseases prior to or on arrival at port of entry customs. After an appropriate inspection has been carried out, a phytosanitary or sanitary certificate can be issued.

A high inspection capacity is essential to cope with the ever-growing volume of trade and travel. Dogs have proved useful for detection of some alien species (Case Study 3.19 "Beagle Brigade Assists in the Search for Forbidden Imports"). X-ray and related equipment is frequently used to inspect travellers' baggage, and its value to detect invasive alien species, such as fruit, seeds and small animals, has improved significantly in recent years. However, this equipment is not installed in many border inspection areas. Some innovative methods to detect live organisms in baggage, such as carbon dioxide detecting machines, are being developed.

In addition to border inspection, searches for pest species can be carried out by on-site inspections. Post entry inspection of plants and animals can be used to check for presence of associated alien species. Certification of the origin of produce in pest-free zones can be used to simplify border procedures in some cases.

Treatment technologies for pathways to prevent bioinvasions

Fumigation is a method frequently used to kill insect infestations in fruits, vegetables, timber, etc. Commodities are treated with gases (e.g. methyl bromide, although this is increasingly being phased out and alternatives sought), at specific atmospheric pressures and temperatures for specific time periods, depending on the commodities and the suspected pests. A common technology to clean grain is the application of carbon dioxide. Other chemical treatment protocols assign use of fluids and dipping procedures.

Temperature treatments involve cold or hot temperatures. Commodities are refrigerated at specific temperatures for a specific number of days or fruits and vegetables are frozen at subzero temperatures with subsequent storage and transportation. Similarly, commodities are dipped into hot water at specific

temperatures for specific time periods. Heat treatment of ballast water using waste energy in the form of high temperatures generated by the engines has been suggested.

Ultra Violet Light (UVL) sterilization of ballast water is a viable and environmentally benign approach. High intensity UVL irradiation is particularly efficient for small organisms. Irradiation is also used to treat commodities.

Another method against organisms in ballast water currently being investigated is the use of **filter** systems aboard ships. These filter systems on the intakes of the ballast water tanks are only likely to be effective against larger organisms, unless some form of pressure filtration system can be developed. This method could prove particularly successful in combination with UV light sterilization, with one method targeting larger organisms and the other smaller ones.

At present mid-ocean **ballast water exchange** remains the primary treatment option recommended for international ship traffic. In most parts of the world ballast water exchange is still on a voluntary basis, but some countries are considering the possibility of a mandatory approach. The major problems involved with ballast water exchange at sea are the lack of ship stability during the exchange process, especially in conditions of high seas, and the lack of efficacy of the method - studies have demonstrated varying effectiveness against different biota found in ballast water. Whereas numbers of individuals in some taxonomic groups were drastically reduced by water ballast exchange, others groups were not significantly affected.

In a significant number of cases a **combination of treatment technologies** is used. When the efficacy of a single method is not satisfactory and the risk of introduction of non-indigenous organisms exceeds an acceptable level, a combination of different methods often leads to success. In some cases an acceptable level of success using a single method can only be reached by applying a specific treatment in such a high dosage that the commodity itself would be damaged, but applying two methods in sublethal dosages sequentially or simultaneously can achieve the security level required.

3.4 Risk assessments

Risk assessment is a tool that can be used to support exclusion of invasive species as well as to assess the potential impact of those that have become established. Risk assessment should be closely tied to risk communication and risk management. Results of risk assessment can be used in decision-making to help determine if action should be taken, and, if so, what kind. Risk assessment can also assist in setting priorities for the best use of time and funds, particularly where there are multiple threats. The risk assessment process and results can be used to build and obtain public support and needed funding for exclusion or eradication.

The risk assessment process is commonly used to rate and rank known or suspected invasive species. The objectives are a prediction of whether or not a species is likely to be invasive and a relative ranking of risk. Entire pathways may also be analysed for risk, and this may be a more efficient procedure where many possible species and vectors are involved. Since financial and other resources are often limited, potential pathways have to be prioritised according to the country's most serious and immediate invasive threats, and managed accordingly (cf. section 2.5.1).

Individual species can also be analysed after they have become established. In this case, ecological modelling and economic analysis may be especially important components of the analysis. Finally, risk assessment may be used to clear species for introduction. The danger here, though, is the problem of false negatives. A certain number of species will be cleared that will later turn out to be invasive.

The risk assessment process commonly begins with the identification of candidate species and pathways. The likelihood of successful introduction is assessed through review of scientific and other literature, expert opinion, and qualitative and/or quantitative analysis. Some of the factors often considered include known invasiveness, likelihood of entry, likelihood of establishment, rate of spread and likely economic and environmental impact. The result is usually some ranking of relative risk, ranging from a simple qualitative rating of "high," "medium" or "low" to a numeric score. Ecological and economic models can also be used to estimate the rate and extent of spread and the potential biological and economic consequences of establishment of a pest or group of pests.

Assessing risk of entry, establishment and spread of potentially invasive species is still a developing field. Assessment schemes have been implemented in only a few countries, and it's too early to reach any conclusions about their success (Case Study 3.20 "Australia's Weed Risk Assessment System"). However, some desirable characteristics to be considered in development of a risk analysis scheme would include the following:

➤ Identifies and utilizes characteristics that are highly correlated with successful introduction, establishment, and spread (conversely, ignores or minimizes the use of those that are trivial);

➤ Uses as few traits as possible while retaining accuracy;

➤ Where possible, uses traits that can be quickly, easily and cheaply determined;

➤ Uses traits that are clearly measurable, wherever possible;

➤ Utilizes open-ended and non-linear probabilities, if appropriate;

➤ Provides for interaction between factors (i.e. a change in the probability of one thing happening might increase or decrease the probability of other things happening);

➤ Assumes that any given species will be eventually distributed throughout its available range unless stopped by a significant physical barrier;

➤ Can be done at reasonable cost;

➤ Has a valid scientific and logical basis;

➤ Effectively discriminates as to level of risk;

➤ Provides a realistic estimate (or range of estimates) of economic impact;

➤ For already introduced species, provides an estimate of the feasibility and cost of eradication or control;

➤ Can be tested for validity (against a different population than that used to devise the system);

➤ Is documented and produces documented results;

➤ Is transparent and open to public review and comment.

Only one factor has a consistently high correlation with invasiveness: whether or not the species is invasive elsewhere. A match of climate and habitat also helps in predicting invasiveness, but many species are known to expand to other habitat types once outside their native range. Characteristics of the species itself in its native range are less accurate predictors (cf. Case Study 3.23 "Invasiveness Cannot Be Reliably Predicted"). These include reproductive and dispersal mechanisms, tolerance to environmental factors such as shade or salinity, life form or habit (e.g. a climbing vine or an aquatic species), and adaptive mechanisms such as the ability of a plant to fix nitrogen. Once a species becomes established, though, these characteristics are more important, as the need at this point is to predict rate and extent of spread. Other factors to be considered in assessing the likelihood of entry include pathways the organism might take, vectors that might help transfer the organism and general preventive measures as well as those that can be employed specifically against the organism.

Risk communication is the communication of risk assessment results so that they are clearly understood and rational decisions can be made. Risk assessment results must be communicated both to decision-makers and the public that must support decisions and the resultant actions. It is important that the process be open and honest, and that public input or participation be solicited at appropriate points throughout the process. Public understanding, acceptance and support are usually essential for effective action against a pest species. Deliberate introduction of potentially risky species should only be done with the informed consent of the public.

Risk management deals with what to do about identified risks. Management of identified risks begins by setting the results of the risk assessment process and other analysis against available options through a decision-making process. The objective is to develop a strategy and plan of action. There are usually a number of risks and limited resources to deal with them. For established pests, several management options (ranging from doing nothing to exclusion to eradication to control measures) are available (see also Chapter 5). Control options include physical, biological and chemical control, each with advantages and disadvantages. Various techniques such as the use of probability theory can be used to support the decision-making process.

Economic and ecological models can be used as part of the assessment and management process to estimate the potential consequences of the establishment of a specific pest or group of pests. Assessment of economic impact, while it usually requires making some assumptions, is highly desirable. The public, decision-makers and legislators understand monetary impacts, costs and benefits while they may not understand the implications of impacts presented solely in ecological terms. Economic analysis of natural resource values is often avoided because it is more difficult than analysis of things that have established market value, such as agricultural crops. However, there are techniques available to make assumptions and obtain agreement on resource values. This is not to say that economic factors should necessarily predominate in decision-making. Other factors, which are largely intangible and do not lend themselves to economic analysis, also need to be considered. These include such things as the cumulative impact of a number of pest species, the irreversibility of the decision to introduce a species, aesthetic and spiritual values and the impact on threatened or endangered species. A very long-term view is also needed, given the slow spread of some invasive species, which may give a misleading result if high discount rates are used in economic analysis. Nevertheless, economic analysis, utilising the best information and assumptions available, is a powerful tool for deciding whether or not to exclude a species, to take action on an incursion, for prioritising actions when dealing with multiple risks and for obtaining needed funding.

Risks can be managed both on a species-specific basis and on a larger scale. When a new invasive species becomes established, a rapid assessment of the risk of spread and the ecological and economic consequences is needed to establish whether or not control or eradication measures are needed. Likewise, a species-specific assessment can be performed for species proposed for introduction. On the other hand, it is often desirable to assess a range of known and potential threats and their potential introduction pathways and develop an overall management strategy. This can aid in the effective and efficient allocation of resources over time to deal with a variety of known and unknown threats.

Risk assessment, because it is a disciplined process, can reduce the amount of subjective judgement involved (although assumptions must still be made - sometimes large ones). It should reduce bias toward such things as particularly charismatic species (Case Study 3.21 "Two Views of the Rainbow Lorikeet in New Zealand"), balance out optimistic and pessimistic approaches, and reduce the use of intuition (which often grossly overestimates or underestimates risk). Since all available knowledge is used, particularly scientific information, formal risk assessments can be better defended to decision-makers, the public, and, if necessary, in court. However, they should still be carried out in an open process and subject to peer and public review.

On the other hand, application of the risk assessment process can be labour-intensive, time-consuming and costly. For example, a risk assessment of importing unprocessed logs from Russia to the United States was estimated to cost US$ 500,000 (Case Study 3.22 "Siberian Timber Imports: Analysis of a Potentially High-Risk Pathway"). However, the potential benefits of exclusion need to be weighed against the cost. In this case, an economic analysis conducted as part of the assessment estimated potential impacts of up to US$ 58 billion from the possible introduction of defoliating insects alone (US Department of Agriculture, 1991).

There are a huge number of potentially invasive species (Case Study 3.23 "Invasiveness Cannot be Reliably Predicted"), and rating even a relatively small percentage of them would take a large effort. Even though risk analysis should be a disciplined process, experts are not without their biases. Lack of knowledge about many species may require many assumptions to be made, leading to lack of confidence in the results. Little may be known about obscure species, and predicting behaviour outside a species' native range is particularly uncertain. Science continues to discover totally new species. Correlation of most characteristics with invasiveness (except for prior evidence of invasiveness) is poor. These factors can lead to the conclusion that a species is potentially invasive when it is not ("false positive") or, more troubling, the conclusion that a species is not invasive when it actually is ("false negative"). Thus, the assessment process and the numeric ratings that are often produced may lead people to put more faith in risk assessment than is justified.

In reality, risk assessment is only one tool and cannot be depended upon exclusively to provide absolute assurance that a species is invasive or innocuous. On the other hand, it provides a logical process for gathering, analysing, synthesising, comparing and communicating information, which can improve the quality of decision-making. Further information on the process can be found through the sources listed in Box 3.1 "Some Pest Risk Assessment Information Sources".

BOX 3.1 Some Pest Risk Assessment Information Sources

References

International Plant Protection Convention (1996) *Guidelines for Pest Risk Analysis.* International Standards For Phytosanitary Measures, 2. Secretariat of the International Plant Protection Convention, Food and Agriculture Organization of the United Nations, Rome, 21 pp. Also available at http://www.fao.org/ag/agp/agpp/pq/default.htm under International Standards for Phytosanitary Measures.

Simberloff, D.; Alexander, M. (1998) Assessing risks from biological introductions (excluding GMOs) for ecological systems. Pp. 147-176 in Calow, P. (ed.) *Handbook of environmental risk assessment and management.* Blackwell, Oxford, UK.

Smith, C.S.; Lonsdale, M.W.; Fortune, J. (1998). Predicting weediness in a quarantine context. *Proceedings of the 6th EWRS Mediterranean Symposium*, (eds. J. Maillet & M.-L. Navas), pp. 33-40. European Weed Research Society, Montpellier.

Tucker, K. C.; and Richardson, D. M. (1995) An expert system for screening potentially invasive alien plants in South African fynbos. *Journal of Environmental Management* **44**, 309-338.

U.S. Department of Agriculture, Forest Service (1991) Pest risk assessment of the importation of larch from Siberia and the Soviet Far East. USDA Miscellaneous Publication No. **1495**.

CAB International / European and Mediterranean Plant Protection Organization (1997) Quarantine pests for Europe. Second Edition. CAB International, Wallingford, Oxon, UK, 1425 pp.

European and Mediterranean Plant Protection Organisation (1993) Guidelines on pest risk analysis. No. 1. Checklist of information required for pest risk analysis (PRA). *Bulletin OEPP/EPPO Bulletin* 23, 191-198.

Internet sites

http://aphisweb.aphis.usda.gov/ppq/weeds/weedsrisk99.html. USDA APHIS Plant Protection and Quarantine: Weed-Initiated Pest Risk Assessment: Guidelines & Template for Qualitative Assessments.

http://www.aphis.usda.gov/ppq/ss/cobra/. USDA-APHIS Plant Protection and Quarantine's Commodity and Biological Risk Assessment (CoBRA) homepage.

http://www.aqis.gov.au/docs/anpolicy/risk.pdf. Handbook of "The AQIS Import Risk Analysis Process" prepared by the Australian Quarantine and Inspection Service; and http://www.aqis.gov.au/docs/plpolicy/wrmanu.htm the AQIS Weed Risk Assessment (WRA) system.

http://www.fao.org/ag/agp/agpp/pq/default.htm. The International Plant Protection Convention – includes the International Standards For Phytosanitary Measures.

http://www.oie.int/eng/en_index.htm Office International des Epizooties – the world organisation for animal health.

http://www.maf.govt.nz/MAFnet/index.htm. Risk analysis prepared by Ministry of Agriculture and Forestry (MAF), New Zealand.

CASE STUDY 3.1   Rosy Wolf snail *Euglandina rosea*, Exterminates Endemic Island Snails

Rosy wolfsnail (*Euglandina rosea*) is a predator of other snails and has been widely used as a biological control agent to try and control various snail pests. This native of Latin America and the South-eastern USA first was introduced to Hawaii in 1955 to combat an exotic agricultural pest, the giant African snail (*Achatina fulica*), and has since been introduced to more than 20 oceanic island groups to try to control this and other snail pests. Some success has been reported against some of the target snail pests, e.g. a 1958-1960 release in Bermuda is reported to have given good control of *Otala lactea*, an introduced snail pest species. There are suggestions that *E. rosea* is quite effective against small individuals of giant African snail, but these have not been quantitatively evaluated and there is no indication that *E. rosea* has controlled *A. fulica* anywhere. What has become clear is that the populations of indigenous snails are very much at risk due to the effectiveness of *E. rosea* as a snail predator.

In Mauritius, 24 of the 106 endemic snail species have become extinct, and on the island of Moorea in French Polynesia, *E. rosea* was a major contributor to the extinction of seven endemic snails in the genus *Partulina*. In most or all island groups where *E. rosea* has been introduced, similar impact has been reported.

Because of Hawaii's isolation and its highly dissected topography, nearly 800 non-marine snail species have evolved there - a textbook case of evolutionary diversification. Because Hawaii's indigenous land and freshwater snail species evolved with few predators, they lack physical or behavioural defences against the alien *E. rosea*. On the island of Oahu, the alien snail is responsible for the loss of most of the 15 to 20 endemic species of *Achatinella* snails that have vanished over the past four decades. This catapulted the entire genus *Achatinella* onto the USA endangered species list. Similarly, some 50% of the species in the closely related genus *Partulina* - found on Molokai, Maui, Oahu, Lanai, and the Big Island of Hawaii-also have been devastated.

Depredation by *E. rosea*, rats, and human shell collectors, along with large-scale loss of forest habitat from logging, farming, urbanization, and feral animal disturbance, has already eliminated 50 to 75 % of the Hawaiian land snails. *E. rosea* is a critical factor in this process that is helping to seal their fate. To protect remaining native snail populations, conservationists are working to prevent the further spread of *E. rosea* into uninfested areas. They also have developed a toxic bait for the invader using the bodies of another pest snail of the genus *Pomacea*. These activities are complemented by efforts to protect undisturbed, intact forests that serve as snail havens and by the establishment of captive breeding colonies of endangered snail species.

This is a good example of what can happen when biological control is undertaken without a critical evaluation of the risks involved. It was well known that *E. rosea* feeds on a wide range of snail species, and so any risk analysis of its introduction should identify that indigenous and endemic snail species would be at risk. Whether this would deter a nation facing a massive invasion of giant African snail from introducing *E. rosea* is another matter – as recently as the 1990s new introductions were being considered in spite of the predator's known track record.

Various sources including Stein, Bruce A. and Stephanie R. Flack, eds. 1996. America's Least Wanted: Alien Species Invasions of U.S. Ecosystems. *The Nature Conservancy, Arlington, Virginia, available through http://www.tnc.org/*

CASE STUDY 3.2 How Africanised Honey Bees Came to the Americas

Africanised honey bees have spread through most of the Americas partly because of their tendency to move more frequently than other honeybees. Their biggest move, however, crossing the Atlantic from Africa to Brazil, was done with human help.

By the 20th century, European honeybees had been imported into South America. These honey bees from colder and drier climates never adapted well to the hot, wet and humid conditions of Brazil. Beekeepers began investigating how they might breed a bee better suited to their environment. Some thought the answer might be found in the tropical zone of Africa. There were reports of beekeepers in South Africa getting remarkable production from indigenous honeybees. African people had been obtaining honey from these wild honeybees for many centuries, and while they knew how furious the insects could get, they had also developed ways to avoid attack.

In 1956 a prominent Brazilian geneticist Warwick Kerr, an expert on Brazil's native stingless bees and familiar with bee breeding and apiculture, was asked by the Brazilian Agriculture Ministry if he could obtain some African honey bee queens and bring them back for breeding experiments. Kerr thought he could utilize African stock to produce a new breed of honey bees, which would be less defensive than the wild African bees but which would be more productive than European honey bees in Brazil's tropical setting. He returned to Brazil with 63 live queens from South Africa, which were taken to a quarantine area at an agricultural research station. By interbreeding the queens through artificial insemination with European drones, Kerr produced first generation hybrids. At this stage 29 Africanised honeybee colonies were maintained in hive boxes equipped with queen excluders (a device put over the hive entrance with holes too small to allow the queen to escape but large enough for the workers to pass through, so that the normal activity of the hive is maintained while the danger of swarming is eliminated).

In October of 1957, however, according to the story that Warwick Kerr has told countless times, a local beekeeper wandered by, noticed the queen excluders and removed them. Such excluders are normally only used in the time before queens begin laying eggs and it is possible that the fellow was just trying to be helpful. In any case, as the story goes, the removal of the excluders accidentally released 26 Africanised honeybee queens with small swarms into the lush forest nearby. By the time Kerr learned of the accident, there was no way of figuring out where the bees had gone. He continued his work with the remaining Africanised honeybees and hybrid queens thinking that perhaps the escaped bees would either perish in the wild or mate with European honeybees and eventually lose their African characteristics.

Within a few years, however, the researchers at Rio Claro began getting reports from surrounding rural areas of feral bees furiously attacking farm animals and even humans. Many poor Brazilian farmers suffered livestock losses, and, eventually, there were human fatalities as well. By the early 1960s, it was clear that a rapid expansion had occurred among feral bee colonies and that the Africanised honey bees were moving quickly into other parts of the country. The rest is history.

Edited from The University of Arizona Africanised Honey Bee Education Project, Information Sheet 15: Africanised Honey Bees: Historical Perspective at
http://ag.arizona.edu/pubs/insects/ahb/inf15.html

CASE STUDY 3.3   The Precautionary Principle

The precautionary principle or approach, appearing in numerous international treaties and declarations, is in essence quite simple and straightforward. Where an activity raises threats of harm to the environment or human health, precautionary measures should be taken even if certain cause and effect relationships are not established scientifically. A common-sense phrasing is "an ounce of prevention is worth a pound of cure."

The precautionary principle, while subject to varying interpretations and having over 12 different definitions in international treaties and declarations, is fast becoming a fundamental principle of international environmental law. In the late 1980s and 1990s, the principle was quickly adopted into numerous multilateral treaties and international declarations, including the 1992 Convention on Biological Diversity.

Variations in terminology have emerged reflecting the considerable controversy surrounding the principle. To avoid the more extreme versions of the precautionary principle that press for absolute environmental protection, some prefer to use the term precautionary approach rather than precautionary principle. Some authors have labelled eco-centric positions as "stronger versions" in contrast to referring to more utilitarian articulations as "weaker versions."

Although variations may occur in threshold criteria and the stringency of environmental control measures, a conceptual core may be delineated. James Cameron, Director of the Foundation for International Environmental Law and Development (FIELD) at King's College of London, has stated the core as follows:

The precautionary principle stipulates that where the environmental risks being run by regulatory inaction are in some way a) uncertain but b) non-negligible, regulatory inaction is unjustified.

A number of core elements or key directions have also been identified. They include

➤   being proactive, a willingness to take action in advance of formal scientific proof;

➤   cost-effectiveness of action, that is, some consideration of proportionality of costs;

➤   providing ecological margins of error;

➤   intrinsic value of non-human entities;

➤   a shift in the onus of proof to those who propose change;

➤   concern with future generations; and

➤   paying for ecological debts through strict/absolute liability regimes.

Edited from a paper prepared by Dr. David VanderZwaag, Director of the Marine and Environmental Law Program (MELP) at Dalhousie Law School, for Environment Canada as part of a review of the Canadian Environmental Protection Act (CEPA) *available at http://www.ec.gc.ca/cepa/ip18/e18_01.html#J13.*

CASE STUDY 3.4   The Impact of Zebra Mussel on Ecosystems

A recent invader to North America, the zebra mussel, *Dreissena polymorpha*, is overwhelming aquatic systems throughout the Great Lakes and Mississippi basins, and could lead to a massive extinction of native freshwater mussels. This mollusc is causing large-scale ecosystem changes and hastening the decline of native freshwater mussels, the USA's most threatened animal group. Its economic impact is as great - the U.S. Fish and Wildlife Service expects it to cause US$5 billion in damages by the year 2002.

Native to the Caspian and Black Seas, the tiny striped-shelled mussel was discovered in North America in 1988. Marine biologists believe it arrived in ballast water of transatlantic ship traffic - the ballast water was discharged, including mussel larvae into Lake St. Clair, between Lakes Huron and Erie. Since then, the prolific creature has spread rapidly throughout lakes and waterways of the eastern United States and Canada, from the Great Lakes through the Mississippi River drainage. It remains unchecked by predators or parasites. Human-made canals and recreational boat traffic enhance the spread of the invader.

Because they cement themselves to any and all submerged hard surfaces, zebra mussels exact a heavy economic and ecological toll. They feed on phytoplankton, outcompeting zooplankton for this essential food and disrupting natural food webs. They also adhere to the shells of freshwater mussels - sometimes in numbers exceeding 10,000 zebra mussels to a single native mussel - thereby interfering with the natives' feeding, growth, movement, respiration, and reproduction. Native mollusc populations tend to crash within four years of zebra mussel colonization.

Researchers predict that zebra mussel invasions of the Mississippi River basin will reduce its native mussel species by as much as 50% within a decade. Because native mussels play an important role in nutrient cycling and sediment mixing, this could seriously affect the ecology of the Mississippi River system. The basin contains more endemic species of freshwater mussels than any other river system in the world. Consequently, the loss of its native mussel life on a scale similar to that already seen in the Great Lakes could result in the extinction of up to 140 species.

Since the zebra mussel's arrival, a number of institutions, including the Great Lakes Sea Grant Network and the US Geological Survey, have developed public education and mussel monitoring efforts. These partners recommend a number of precautions individuals can take to prevent new zebra mussel introductions: remove attached vegetation and wash boats or trailers before moving them to new lakes or rivers; flush engine cooling systems, bilging areas, and live wells with tap water; leave unused bait and bait bucket water behind; and inspect boat hulls for signs of zebra mussels before relocating the vessels.

Edited from: Stein, B. A.; Flack, S.R. (eds.) (1996) America's Least Wanted: Alien Species Invasions of U.S. Ecosystems. *The Nature Conservancy, Arlington, Virginia. Available through* http://www.tnc.org/

CASE STUDY 3.5 Asian Longhorned Beetle, a Threat to North American Forests

Anoplophora glabripennis, the Asian longhorned beetle (ALHB) has been discovered in the United States. ALHB probably travelled to the United States inside solid wood packing material from China, and the beetle has been intercepted at ports and found in warehouses throughout the United States. It has been found attacking trees and has been controlled at these locations.

The insect is 1 1/4" long, coal black with irregular white spots on its back. It has 2" long black antennae with white rings. The females chew oval, darkened notches in the bark of trees, into which they deposit their eggs. After the eggs hatch, the larvae bore into the tree, feeding on the wood. The larvae may feed on the heartwood of the tree all winter. Tunnelling by beetle larvae girdles tree stems and branches. When mature, the beetles burrow out of the tree in late spring or summer, leaving a 3/8" hole where they exit. Adult beetles then feed on the bark and leaves of trees. Repeated attacks lead to dieback of the tree crown and, eventually, death of the tree.

The insect is a serious pest in China where it has few natural enemies; in North America no natural enemies have been recorded as yet. If this insect becomes established in the environment, it could destroy millions of acres of America's treasured hardwoods. In the USA the beetle prefers maple species (*Acer* spp.), including box elder, Norway, red, silver, sugar and sycamore maples. Maples are not only a dominant tree species in the northeastern part of the USA but a US$40 million industry producing maple sugar. It also attacks many different hardwood trees, including horse chestnut, mulberry, black locust, elms, birches, willows, poplars and green ash.

Currently, the only effective means to eliminate ALHB is to remove infested trees and destroy them and the eggs and larvae within them by chipping or burning. Early detection of infestations and rapid treatment response are crucial to successful eradication of the beetle. To prevent further spread of the insect, quarantines are established to avoid transporting infested trees and branches from the area.

In 1996, State and Federal Governments spent more than US$4 million on a suppression programme in New York City and Amityville, NY, neighbourhoods, which is not believed to have eradicated the beetle.

Compiled from the USDA Forest Service Pest Alert at http://willow.ncfes.umn.edu/albpestalert/, *the USDA-APHIS site at* http://www.aphis.usda.gov/oa/pubs/fsalb.html *and the Illinois Department of Agriculture site at* http://www.agr.state.il.us/beetle.html.

CASE STUDY 3.6 The Threat of South American Leaf Blight to Rubber in Malaysia

Rubber (*Hevea brasiliensis*) is a plant of South American origin, grown extensively in South-east Asia, especially in Malaysia, for the production of natural rubber. As an introduced crop it is relatively free of pests and diseases, and specifically no significant pests or diseases have accompanied this crop from its area of origin in the Amazon.

The most damaging disease of rubber in South America is South American Leaf Blight (SALB, *Dothidella ulei*), and it is so virulent that in practice the commercial planting of rubber in the continent is not viable. Malaysia, the main South-east Asian rubber producer, has a substantial programme aimed at preventing the entry of this disease, which R.E. Schultes, Director Emeritus of the Harvard Botanical Museum, has suggested would run through the Asian plantations in five years, reducing yields, killing trees, and compromising the entire industry. The total area planted with rubber in Malaysia in 1997 was 1.564 million ha mostly by small holders. The total rubber production for Malaysia in 1999 was 0.9 million tonnes and the export value derived was RM 3,115 million or 2% of the total Malaysian exports.

The first and principle line of defence in the region is prevention. Quarantine regulations in Malaysia, Thailand and some other natural rubber producing countries have been strengthened to prevent an accidental introduction of leaf blight into these countries. The importation of rubber tree planting materials directly from the American tropics is prohibited except for research purposes. Airport posters are used to alert passengers and relevant national and industry research institutions and universities have been sensitised. Passengers coming from tropical South American countries are requested to stopover in another country en route for at least two days and those arriving on direct flights are required to complete plant quarantine declaration cards and upon their arrival they (together with their baggage) are subjected to quarantine treatments e.g. to shower and change their clothing, and exposure of their baggage to ultra violet light irradiation).

Investment in early warning systems is quite limited by comparison. Staff of the Rubber Research Institute of Malaysia together with staff of the Department of Agriculture carry out regular surveys of rubber diseases every two to three years. These are intended to identify centres of infestation by indigenous disease problems, so that recommendations of suitable clones can be made for different areas. However, whilst doing this, a watch is kept for symptoms of other diseases, especially South American Leaf Blight.

A contingency plan has been prepared in the event of leaf blight being found in Malaysia, but by that stage, it may well be too late for the South-east Asian rubber industry.

Prepared by Soetikno Sastroutomo, CAB International South-East Asia Regional Centre, Malaysia Agricultural Research & Development Institute, P.O. Box 210, 43409 UPM Serdang, Malaysia; searc@cabi.org

CASE STUDY 3.7   Anecdotes about Entry Pathways

Many observations and speculations about invasive species, their mode of entry and impact are recorded in anecdotal form.

For example, Lucas Bridges writes about missionary work in Tierra del Fuego (Argentina and Chile) where he was born and brought up in a missionary family at the end of the nineteenth century and early twentieth century. He describes how charitable people in England would regularly send large consignments of clothing for distribution amongst the local Yahgan people, and that these were "generally somewhat shop-soiled". He goes on to add a footnote: "It is of interest to put on record that a fine lawn grass, not indigenous to the country, made its appearance and spread rapidly around the Yahgan settlements. Father was strongly of the opinion that the seed had been brought adhering to the soles of used tennis shoes."

In the same work Lucas describes how his father brought some rabbits from the Falkland Islands, and used this stock to colonize small islands in the area to provide "welcome food for the natives and any shipwrecked crews who might be stranded there". The rabbits were not allowed to escape on the main island (Tierra del Fuego Island) or any of the larger islands lest they become pests to farmers. On those islands with good sandy soil and bush, the rabbits thrived.

European rabbits (as well as North American beaver and Arctic Reindeer) are now well established in Tierra del Fuego and are considered to have devastated the local flora.

Source: Lucas Bridges (1948) Uttermost Part of the Earth, reprinted 1987 by Century Hutchinson, London. See also
http://www.gorp.com/gorp/location/latamer/argentin/tierra.htm

CASE STUDY 3.8   Spread of a Biological Control Agent, *Cactoblastis cactorum*, in the Caribbean

The very successful introduction of *Cactoblastis cactorum* moth from Argentina into Australia in the 1920s for the biological control of *Opuntia* cactus is a flagship programme for weed biological control, since repeated in South Africa and other countries. In 1957 this species was introduced into the Caribbean island of Nevis, resulting in what was considered an outstanding control of the target alien weedy *Opuntia* species. Subsequently, the moth spread the two miles to the island of St. Kitts, and it was introduced into the Cayman Islands, Antigua and Montserrat.

In recent years, the moth has spread more widely in the Caribbean and in 1989 reached Florida, and in 2000 was reported from the Yucatan Peninsula, Mexico. The mechanism of introduction to the mainland is unknown, but is assumed to have been from the Caribbean. In the case of Florida it probably came in infested *Opuntia* sp. carried by human activity from the Dominican Republic.

In Florida it is now causing concern because of attacks on indigenous *Opuntia* species. In particular, the endemic semaphore cactus (*O. spinosissima*), brought to the edge of extinction by habitat destruction may be pushed to extinction in the wild. The threat to indigenous *Opuntia* spp. in Mexico has yet to be evaluated, but *Opuntia* cultivars are a significant local crop, used for their pads and fruits.

It is not straightforward to identify responsibility in this situation. Some might blame the original biological control introduction into Nevis, others the breakdown in quarantine that allowed *C. cactorum* to spread to areas where it is now considered a pest.

It is worth noting that it is not the predictability of impact on non-target organisms that is at the root of this – *C. cactorum* was known to attack a range of *Opuntia* species. Rather it is the decision-making process and the effectiveness of quarantine that can be questioned. Nevis may have largely forgotten about the biological control of *Opuntia* species 40 years ago, but if reminded would feel that the correct decision was made. The USA and Mexico and the Caribbean island countries do not consult each other about what biological control agents will be released. What could with some justification be highlighted is the irreversibility of biological control and that the values and concerns of society change over time, so that yesterday's correct decision can look questionable today. This of course is true of any irreversible decision that a government makes.

Prepared by Matthew Cock, CABI Bioscience Switzerland Centre, 1 Rue des Grillons, CH-2800 Delémont, Switzerland. www.cabi.org/bioscience/switz.htm.

CASE STUDY 3.9 Release of Exotic Fish by Aquarium Hobbyists – the USA Experience

Most fishes available for sale in pet shops are exotic and are imported into the USA predominantly from Central and South America, Africa, and South-east Asia. Each year, over 2000 species, representing nearly 150 million exotic freshwater and marine fishes, are imported into the USA for use in the aquarium trade. Unfortunately, some exotic fishes are released into the wild each year. Hobbyists may not be able to take their fish with them when they move, or they simply may lose interest in maintaining an aquarium. Fish may also be released if they outgrow the aquarium or if they appear to be in poor health.

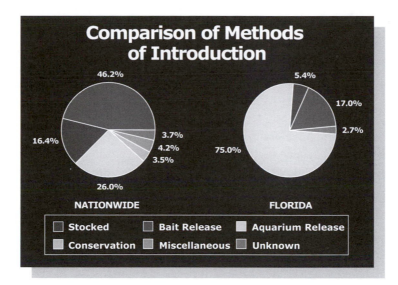

Currently, at least 185 different species of exotic fishes have been caught in open waters of the USA, and 75 of these are known to have established breeding populations. Over half of these introductions are due to the release or escape of aquarium fishes. Because many of these fishes are native to tropical regions of the world, their thermal requirements usually prevent them from surviving in temperate areas. In the USA, therefore, most introduced fishes have become established in Florida, Texas, and the South-west USA. Examples include a number of cichlids, such as the oscar, Jack Dempsey, jewelfish, convict cichlid, Midas cichlid, and spotted tilapia; and livebearers, such as swordtails, platies and mollies, and armoured catfishes. The goldfish, a native of China, is one of the few examples of a temperate aquarium species that is established throughout the USA.

Instead of subjecting the fish to potentially harmful environmental conditions or risking potential ecological problems by releasing it, there are alternative means for disposing of unwanted pet fish:

➤ Return it to a local pet shop for resale or trade.

➤ Give it to another hobbyist, an aquarium in a professional office, a museum, or to a public aquarium or zoological park.

➤ Donate it to a public institution, such as a school, nursing home, hospital, or prison.

If these options are not available, rather than release fish into the wild, they should be "put to sleep", e.g. by placing the fish in a container of water and putting it into the freezer. Because cold temperature is a natural anaesthetic to tropical fishes, this is considered a very humane method of euthanasia.

Edited from the <u>U. S. Department of the Interior Geological Survey</u> Non-indigenous Aquatic Species website "Problems with the Release of Exotic Fish " at: http://nas.er.usgs.gov/fishes/dont_rel.htm through http://nas.er.usgs.gov/

CASE STUDY 3.10 The Introduction of Parthenium Weed into Ethiopia

Parthenium weed, *Parthenium hysterophorus* (Asteraceae), is an annual herb with a deep taproot and an erect stem that becomes woody with age. As it matures, the plant develops many branches and may eventually reach a height of two metres. Small creamy white flowers occur on the tips of the numerous stems, each flower containing 4-5 seeds.

Parthenium weed is a native of subtropical areas in South and North America. As an introduced species, it vigorously colonizes weak pastures with sparse ground cover. It will readily colonize arable land, disturbed, bare areas along roadsides and heavily grazed pasture. It is also a health problem as contact with the plant or the pollen can cause serious allergic reactions such as dermatitis and hay fever.

As with most weeds, prevention is much cheaper and easier than cure. *P. hysterophorus* seeds can spread via water, vehicles, machinery, stock, feral and native animals and in feed and seed. Vehicles and implements, especially earthmoving machinery, passing through parthenium weed infested areas should be washed down with water. The wash down procedure should be confined to only one area, so that any plants that establish from dislodged seed can be destroyed before they set seed. Extreme caution should be taken when moving cattle from infested to clean areas.

Biological control has been implemented in Queensland, Australia, and so far, nine species of insect and a rust pathogen have been introduced to control parthenium weed. The combined effects of biological control agents has reduced the density and vigour of parthenium weed and increased grass production.

Parthenium weed was first reported from Ethiopia at Dire-Dawa, Harerge, eastern Ethiopia in 1988. A second major centre of infestation was subsequently found near Dese, Welo, north-eastern Ethiopia. Both are major food-aid distribution centres, and there is a strong implication that parthenium weed seeds were imported from subtropical North America as a contaminant of grain food aid during the 1980s famine, and distributed with the grain.

By 1999, parthenium weed was widespread in eastern Ethiopia, close to Addis Abeba, and reported to be spreading into western Ethiopia. The Awash National Park and the Yangudi Rasa National Park are immediately at risk, as the weed spreads in a series of small to large jumps with the accidental human assistance.

As an exotic invasive weed, *P. hysterophorus* can be expected to continue to expand its range until all suitable habitats are occupied. Efforts to contain it will at best delay this process. Impact on the environment, agriculture and human health will increase, and as the human population becomes sensitised, the medical effects are likely to escalate. Parthenium weed already has a local name, which translates as "no-crop". Since Ethiopia has suffered from famine as much as any country in the world in recent decades, this does not augur well.

Prepared by Matthew Cock, CABI Bioscience Switzerland Centre, 1 Rue des Grillons, CH-2800 Delémont, Switzerland. www.cabi.org/bioscience/switz.htm.
Information on Parthenium edited from Queensland Department of Natural Resources Pest Facts at http://www.dnr.qld.gov.au/resourcenet/fact_sheets/pdf_files/pp2.pdf. See the Centre for Tropical Pest Management's Parthenium site (http://www.ctpm.uq.edu.au/parthenium/parthenium.html) for more information on the weed.

CASE STUDY 3.11   Long-distance Spread of *Miconia calvescens* to Remote Islands of French Polynesia

The alien tree *Miconia calvescens* is a dominant plant invader on the tropical islands of Tahiti, Moorea and Raiatea (Society Islands) where it was intentionally introduced as an ornamental plant.   Amongst the biological characteristics explaining the striking success of this invasive species is a large soil seed bank (up to 50,000 seeds per square metre) and the ability of the seeds to remain viable in the ground for at least six years.

Despite an active research and information programme to control *M. calvescens* (see Case Study 4.6 "Public Awareness and Early Detection of *Miconia calvescens* in French Polynesia"), *M. calvescens* seedlings were recently discovered in remote islands of French Polynesia, 700-1400 km from the Society Islands. Isolated *M. calvescens* plants were found on Rurutu and Rapa (Austral Islands) near water-tanks (reservoirs) built with gravel and soil imported from Tahiti; small populations of *M. calvescens* seedlings were spotted in Nuku Hiva and Fatu Hiva (Marquesas) in 1997, on road sides and in gulches below where road works were carried out using bulldozers from Tahiti; in 1990 *M. calvescens* seedlings were found on Huahine (Society Islands) in the Fare Harbour, growing on a pile of imported gravel and soil.

Accidental introduction of *M. calvescens* through the transportation of contaminated gravel and soil and dirty machinery (bulldozers, tractors) for construction works, is now considered to be the main cause of *M. calvescens* long-distance spread in French Polynesia. In 1999, as recommended by the Inter-Ministerial Committee to Control Miconia and other invasive plants (a committee created in 1998), the Government of French Polynesia and the High-Commissioner of France wrote official letters to contractors for public works, requesting them to clean their vehicles as a quarantine strategy before landing on remote islands

Transportation of potted plants between islands is strictly forbidden in French Polynesia, but illegal introduction of potted plants, which may contain *M. calvescens* infected soil, still occurs. The dispersal of *M. calvescens* by local pig-hunters or by foreign tourists (especially mountain hikers, and biologists!) with muddy shoes coming from Tahiti and Moorea may also be a threat to the remote - and still pristine - high volcanic islands of French Polynesia.

Prepared by Jean-Yves Meyer, Délégation à la Recherche, B.P. 20981 Papeete, Tahiti, French Polynesia. E-mail Jean-Yves.Meyer@sante.gov.pf

CASE STUDY 3.12   The Australian Defence Force is Involved in Keeping Alien Species Out

Finding and cleaning every tiny grass seed from an M113 armoured personnel carrier sounds like a tough job. Cleaning the same seeds off 1000 army vehicles — everything from trucks to front end loaders and water tankers — is even tougher. Then cleaning off every trace of soil, every piece of foliage, insect and egg off 10,000 pallets of army equipment — everything from generators to tents and refrigerators — is stretching the limits of probability. But it had to be done in Dili, East Timor, before 5000 Australian peace keeping soldiers and all their vehicles and equipment could return to Australia.

The likelihood of seeds and plant matter being spread by direct contact with military equipment is high. Weeds and seeds can spread as contaminants in soil stuck to vehicles, machinery, radiators, cuts in tyres, equipment, camouflage netting and personal equipment. Some seeds are light and windborne and are easily trapped in radiator grilles, equipment brackets and other small areas. Soil generally collects around the wheels and tracks of vehicles, but also on boots, personal equipment clothing, tents, packaging boxes and tent poles.

The job of checking that all the vehicles and the equipment (and the troops themselves) were not carrying pests and diseases into Australia fell to the Australian Quarantine and Inspection Service (AQIS). The job of cleaning all that equipment to AQIS standards fell to the Australian Defence Force.

Captain Kevin Hall had the task to devise the washing and inspection procedures, to comply with the AQIS quarantine requirements. He developed an illustrated 160-page manual, which became the bible for the major cleaning operation in Dili that had up to 300 staff operating 20 wash stations 18 hours every day for three months.

This 'how to' clean up manual covers everything from how to clean soil out of the tyres of graders to where insects can lodge in an Unimog. It has photographs of all the Army vehicles and equipment with diagrams on how and where to clean them. It lists the equipment needs — from high-pressure water and air hoses to vacuum cleaners, brushes and even dustpans. All the necessary techniques were developed for the task and documented in the manual, which establishes guidelines that AQIS and the military could use not only for the East Timor operation, but also for future operations.

Captain Hall recently received a 2000 National Quarantine Award from AQIS in recognition of his efforts.

Edited from a media release of the Australian Quarantine and Inspection Service, Department of Agriculture, Fisheries and Forestry, 23rd May 2000, available through http://www.aqis.gov.au/

CASE STUDY 3.13 Hitchhikers Moved with Marine Baitworms and Their Packing Material

In different parts of the world, various species of marine worms are harvested, packed and shipped alive via air transport to other regions for use in sport (recreational) fishing. The worms, the algae that are used in some cases as packing material, and other organisms living on or in the worms and algae are then frequently released into coastal waters in these new regions, where some may become established. The best documented of these pathways involves the shipment of worms from the Atlantic to the Pacific coasts of the United States.

The clam worm (also called the sand worm or pile worm) *Nereis virens* and the bloodworm *Glycera dibranchiata* are dug from intertidal muds on the coast of Maine and shipped in boxes of 125 or 250 worms to other states on the USA Atlantic coast, to California on the US Pacific coast, to France and Italy in Europe, and possibly to other continents. These worms are packed in the intertidal fucoid seaweed *Ascophyllum nodosum*, which may incidentally contain mussels, clams, snails and snail eggs, crabs, amphipods, isopods, copepods, mites, annelid worms, other seaweeds and other organisms, estimated to include hundreds of thousands of individuals per year of some Atlantic species shipped to the Pacific coast. In surveys in California, about a third of recreational fishers reported that they dispose of the seaweed and any unused baitworms in coastal waters or deposit them in intertidal areas (Lau, 1995). At least three species established on the Pacific Coast were possibly or probably introduced via this pathway: the snail *Littorina saxatilis* (Carlton & Cohen 1998), the seaweed *Codium fragile tomentosoides* (native to Asia, but introduced to the Atlantic by 1900), and the green crab *Carcinus maenas*. In the decade since its arrival, this crab has spread from California to southern Canada, and has raised concerns about harmful impacts both on ecosystems and commercial shellfisheries (Cohen *et al.*, 1995).

References

Carlton, J.T.; Cohen, A.N. (1998) Periwinkle's progress: the Atlantic snail Littorina saxatilis (Mollusca: Gastropoda) establishes a colony on a Pacific shore. Veliger 41, 333-338.

Cohen, A. N.; Carlton, J.T.; Fountain, M.C. (1995) Introduction, dispersal and potential impacts of the green crab Carcinus maenas in San Francisco Bay. Marine Biology 122, 225-237.

Lau, W. (1995) Importation of baitworms and shipping seaweed: vectors for introduced species? Pages 21-38. In: Sloan, D. & D. Kelso (eds.), Environmental issues: from a local to a global perspective. Environmental Sciences Senior Seminar, University of California, Berkeley.

CASE STUDY 3.14 Spread of the Brown Tree Snake in the Pacific Region

The brown tree snake (*Boiga irregularis*) is an introduced species on Guam that has become a serious pest. The snake probably arrived on snake-free Guam when military equipment was moved onto Guam immediately after World War II. The first sightings were inland from the seaport in the early 1950s. Snakes became conspicuous throughout central Guam by the 1960s, and by 1968, they had probably dispersed throughout the island.

In the absence of natural population controls and with vulnerable prey on Guam, the snakes have now become an exceptionally common pest causing major ecological and economic problems on the island. Up to 13,000 snakes per square mile may occur in some forested areas of Guam. The snakes feed on a wide variety of animals including lizards, birds, and small mammals as well as bird and reptile eggs. The brown tree snake has virtually wiped out the native forest birds of Guam. Twelve species of birds, some found nowhere else, have disappeared from the island, and several others persist in precariously low numbers close to extinction. Of the 12 native species of lizard, nine are expected to become extinct.

Snakes crawling on electrical lines frequently cause power outages and damage electrical lines. The snakes cause about 86 power outages a year (every 4th to 5th day!) with a conservatively estimated cost of US$1 million/year. The power interruptions cause a multitude of problems ranging from food spoilage to computer failures.

The brown tree snake is aggressive when threatened. It will often raise the anterior body in a striking position, flatten the head and neck to appear larger, and attempt to bite as it lunges forward. Adults can reach lengths of 8 feet and weights of 5 pounds. The brown tree snake is a mildly venomous species that kills its prey by chewing to inject the venom, since the venomous-injecting teeth are in the rear of the mouth (opisthoglyph). It is not known to be fatal to humans, but some bitten infants required hospitalisation and intensive care.

Snakes are frequently accidental stowaways in cargo leaving Guam, and unless intercepted may become established on other islands. Economic and ecological problems like those currently present on Guam would be likely to develop if the brown tree snake were to reach other Pacific Islands.

In addition to Guam, brown tree snakes have been sighted on Saipan, Tinian, Rota, Kwajalein, Wake Oahu, Pohnpei, Okinawa, and Diego Garcia. To date, this snake is not known to be established on any of these islands except Guam, but frequent reports of snake sightings on Saipan evidence the presence of snakes on this island.

Travellers, cargo handlers, and Pacific Island residents alike share in the responsibility to protect island environments from this pest species. Careful inspection of materials, cargo, and baggage shipped from or through Guam is necessary to prevent the dispersal of snakes to other islands. With increased awareness and careful inspection of cargo arriving from Guam, it may be possible to prevent the spread of the brown tree snake to other islands.

Edited from "The brown tree snake: a fact sheet for Pacific island residents and travellers" prepared by Thomas H. Fritts, and available at http://www.pwrc.nbs.gov/btree.htm

CASE STUDY 3.15 Monitoring for the Black Striped Mussel in Northern Territory, Australia

An infestation of the exotic black striped mussel, *Mytilopsis* sp. (also known as *Congeria sallei*) was discovered in Darwin marinas late March 1999. Recognising the potential adverse impact on the Australian economy and biodiversity if the bivalve was to become established in Australian waters, the Northern Territory Government (NTG) implemented an immediate and successful containment and eradication programme (see Case Studies 4.13 and 5.23)

As a consequence of *Mytilopsis* sp. being well established in many ports along international yacht cruising routes, the NTG asked for the co-operation of all internationally travelled vessels intending to enter Darwin marinas. Any international vessel, which cannot demonstrate that the hull was anti-fouled in Australia, is required to undergo a hull inspection and treatment of their internal seawater systems. Internationally travelled vessels anti-fouled in Australia, that have remained in Australian waters, need only have their internal seawater systems treated prior to being permitted entry to Darwin marinas.

The Department of Primary Industry and Fisheries, Aquatic Pest Management Team have inspected, treated and cleared a total of 30 visiting international vessels per month. The skippers of the vessels have proven extremely co-operative and commend the active role the Government has taken in preserving our pristine marine environment.

The value of the inspection and treatment protocols has been demonstrated. A vessel requesting entry to a Darwin marina was denied entry on the basis that it had spent the previous six months in Indonesian waters and had not been anti-fouled since its return to Australian waters. The hull of the vessel was found to be clean, however four species of bivalve were found in the strainers of the internal seawater systems. Of the four species of bivalve, two were considered to be similar in nature to *Mytilopsis* sp.: the Asian green mussel (*Perna veridis*) and the bag mussel (*Musculista* sp.). If these two mussel species had gained entry to a Darwin marina and become established in a similar manner to the black striped mussel, it is quite feasible that the events of April 1999 would have repeated themselves in 2000.

Edited from <http://coburg.nt.gov.au/dpif/fisheries/environ/unittext.shtml>.

CASE STUDY 3.16 Transfer of Pathogens and Other Species via Oyster Culture

In *The Ecology of Invasions by Animals and Plants*, Charles Elton referred to oysters as "a kind of sessile sheep, that are moved from pasture to pasture in the sea." For over 150 years, several species of oyster have been transported around the world and planted in large numbers in coastal waters remote from their native regions, to grow there to marketable size. Many organisms have travelled with them: parasites and commensals hidden within oysters; epibionts attached to or living on rough oyster shells or in among clumps of oysters; and oyster predators, pests and other organisms carried in the mud, water and other materials packed with the oysters. By such means, several damaging shellfish diseases and other pests have been spread to various parts of the world.

Among these are the oyster diseases MSX (*Haplosporidium nelsoni*) and microcell disease (*Bonamia ostreae*), apparently transported with oyster shipments into new regions where they have devastated shellfisheries (Farley 1992). Other pests that have travelled with oysters include a flatworm and several species of oyster drills (snails that bore into oysters and other bivalves) that prey on oysters, sponges that grow into and weaken oyster shells, slipper shells and seaweeds that compete with oysters for space, and a copepod that porosities oysters and reduces their marketability (e.g. Chew 1975; Neushul *et al*. 1992). Large numbers of other oyster bed inhabitants that are not specifically pests of oysters have been transferred as well. For example, several non-indigenous organisms that have become established in San Francisco Bay may have arrived with oyster shipments, including one seaweed, three protozoans, five sponges, five hydroids, two anemones, four oligochaete and eight polychaete worms, three opisthobranchs, six snails, two mussels, four clams, an ostracode, a copepod, six amphipods, a crab, a kamptozoan, five bryozoans and five sea squirts (Cohen and Carlton 1995). Some researchers have suggested that oyster shipments could introduce into new regions certain organisms responsible for human illnesses, such as toxic, red tide forming dinoflagellates and novel strains of cholera. And despite their potential economic value, the oysters themselves may become pests: in Australia the Pacific oyster *Crassostrea gigas* is considered a nuisance species because it competes with native oyster species (Furlani 1996).

Prepared by Andrew Cohen, San Francisco Estuary Institute, 1325 South 46th Street, Richmond, CA 94804, USA.

References
Chew, K.K. (1975) The Pacific oyster (Crassostrea gigas) in the west coast of the United States. Pp. 54-80 in Mann, R. (ed.), Exotic Species in Mariculture. MIT Press, Cambridge MA.
Cohen, A.N.; Carlton, J.T. (1995) Nonindigenous Aquatic Species in a United States Estuary: A Case Study of the Biological Invasions of the San Francisco Bay and Delta. U.S. Fish and Wildlife Service, Washington DC.
Farley, C.A. (1992) Mass mortalities and infectious lethal diseases in bivalve molluscs and associations with geographic transfers of populations. Pp. 139-154 in Rosenfeld, A.; Mann, R. (eds.), Dispersal of Living Organisms into Aquatic Ecosystems. Maryland Sea Grant Publication, College Park MD.
Furlani, D.M. (1996) A Guide to the Introduced Marine Species in Australian Waters. Tech. Rep. No. 5, Centre for Research on Introduced Marine Pests, CSIRO Division of Fisheries, Hobart, Tasmania.
Neushul, M.; Amsler, C.D.; Reed, D.C.; Lewis, R.J. (1992) Introduction of marine plants for aquaculture purposes. Pp. 103-135 in Rosenfeld, A.; Mann, R. (eds.), Dispersal of Living Organisms into Aquatic Ecosystems. Maryland Sea Grant Publication, College Park MD.

CASE STUDY 3.17   Japanese Brown Alga Introduced with Oysters

Japanese brown alga, *Sargassum muticum* (Sargassaceae) is a medium to large (2-10 m) yellowish-brown, bushy seaweed native to Japan. It is found in lower intertidal and shallow subtidal waters of quiet bays and lagoons. It colonizes mud and sand flats and seagrass (*Zostera marina*) beds, fastened to solid substrates such as oysters and rocks.

It was introduced on the shells of Japanese oysters (*Crassostrea gigas*) or with oyster spat transplanted to the Pacific coast, and first became established in North America before 1941 in British Columbia. Detached branches, rendered buoyant by air vesicles, are dispersed by water currents and wind drift down the Pacific Coast. As a result, coastal ship traffic may have carried the seaweed as hull fouling into the San Francisco Bay. It is now locally abundant all along the Pacific Coast of North America. It is also introduced to the coasts of Britain, France, the Netherlands and into the Mediterranean Sea.

Sargassum muticum invades habitats normally occupied by eelgrass (*Zostera marina*). Eelgrass beds are important nurseries for many marine species. Their displacement by *S. muticum* could prove detrimental to the ecosystem of the northern Pacific Coast. As it is fast growing and fertile within the first year of its life, *S. muticum* is expected to out-compete local seaweed species in Europe.

Edited from Fact Sheet on Sargassum muticum *by Colette Jacono on the U. S. Department of the Interior Geological Survey Non-indigenous Aquatic Species website at*
http://nas.er.usgs.gov/algae/sa_mutic.html

CASE STUDY 3.18 Sorry, No Free Rides from the Torres Strait

Sunstate Airlines provides a daily service between Cairns in mainland Australia and Horn Island in the Torres Strait for tourists and business people — but there are some passengers the airline will not carry. Through a practical programme, Sunstate makes sure that it does not carry the pests and diseases that are found in the Torres Strait to Cairns. Two would-be passengers that Sunstate will not accept are mosquitoes, which can carry Japanese encephalitis, dengue fever and malaria and fruit flies, which could damage the mainland's orchard crops.

Also present in Papua New Guinea and the Torres Strait — and on Sunstate's banned passenger list — are Asian honeybees, screwworm flies, mango caterpillars, sugar stem borers, citrus canker and Siam weed. All of these could devastate Australia's agriculture industries and some would be major environmental problems.

The airline keeps these unwanted passengers away with a routine that includes providing quarantine information to passengers and regular disinfection of its aircraft. Announcements are made on board and passengers are informed before they leave Horn Island about Australia's quarantine restrictions. Every passenger gets a quarantine message in his or her ticket wallet.

Quarantine information cards are placed in every aircraft seat — and the staff check that every seat has one. Sunstate flight attendants receive training on quarantine regulations — and are tested regularly to ensure they are up to date and informed on quarantine.

According to Sunstate's Operations Manager, all the procedures are simple but effective. "Most of our passengers are regular travellers to and from the Torres Strait, they have become accustomed to the requirements and actually appreciate the fact that Sunstate and AQIS are doing their utmost to safeguard mainland Australia from pests and diseases," he said.

Sunstate Airlines recently received a 2000 National Quarantine Award from Australian Quarantine and Inspection Service in recognition of their efforts.

Edited from a media release of the Australian Quarantine and Inspection Service, Department of Agriculture, Fisheries and Forestry, 23rd May 2000 available through http://www.aqis.gov.au/

CASE STUDY 3.19   Beagle Brigade Assists in the Search for Forbidden Imports

USDA's Beagle Brigade is one facet of APHIS' agricultural quarantine and inspection (AQI) programme. The Beagle Brigade is a group of non-aggressive detector dogs and their human partners. They search travellers' luggage for prohibited fruits, plants, and meat that could harbour harmful plant and animal pests and diseases. These detector dogs work with APHIS inspectors and x-ray technology to prevent the entry of prohibited agricultural items.

In 1996, 66 million people travelled to the United States. In addition, there are many millions of pieces of international mail and countless commercial import and export shipments. As part of the APHIS programme, Plant Protection and Quarantine (PPQ) officers inspect passenger baggage, mail, and cargo in the Federal Inspection Service (FIS) areas at all USA ports of entry. Beagles are used at airports as detector dogs. The Beagle Brigade, which includes the detector dogs and PPQ officers serving as canine handlers, generally works among passengers as they claim their bags.

On average, APHIS PPQ officers make about two million interceptions of illegal agricultural products every year. The Beagle Brigade programme averages around 75,000 seizures of prohibited agricultural products a year.

APHIS selected beagles for use at airports because of their acute sense of smell and their gentle nature with people. Beagles' natural love of food makes them effective detectives and happy to work for treats. APHIS has found that most beagles will remain calm in crowded, noisy locations, such as busy airport baggage-claim areas. These detector dogs are bright, inquisitive, and active hounds whose sense of smell makes them curious wanderers by nature. Beagles have such precisely sensitive scenting ability that they can detect and identify smells so faint or diluted that even high-tech scientific equipment could not measure them.

Humans have an estimated five million scent receptors concentrated in a relatively small area at the back of the nose. By comparison, beagles have an estimated 220 million scent receptors. Not only do beagles have a marvellous ability to detect scents, but also after extensive training, they are good at distinguishing one odour from another and remembering it.

Edited from USDA's Detector Dogs: Protecting American Agriculture at
http://www.aphis.usda.gov/oa/pubs/usdabbb.pdfo

CASE STUDY 3.20 Australia's Weed Risk Assessment System

Australia has recently adopted as part of its new quarantine procedures a weed risk assessment system for assessing plant species for weediness potential prior to introduction to Australia. This system has been endorsed by Environment Australia as well as by a wide range of client groups.

The Weed Risk Assessment (WRA) system is a question-based scoring method. Using the WRA involves answering up to 49 questions on the new species to be imported. The questions include information on the plants: climatic preferences, biological attributes, and reproductive and dispersal method. The WRA uses the responses to the questions to generate a numerical score. The score is used to determine an outcome: accept, reject or further evaluate for the species. The WRA also makes a prediction as to whether a species may be a weed of agriculture or the environment.

In one analysis, the WRA was found to be more decisive than the comparable methods, giving more than 80% accurate results in identifying weeds. However, accuracy depends somewhat on sources used to assess weed status, and the WRA was also rather inaccurate at predicting weeds among the Poaceae and Fabaceae.

AQIS, the agency that takes action to regulate importation of plants, has now formally adopted the system for assessing all new plant imports. To facilitate the process of assessment, information is being requested from prospective importers and a questionnaire has been developed. A package is currently being developed which will allow importers with the necessary expertise, or registered consultants, to conduct pre-entry assessments with the system before lodging an application to import.

Edited from Smith, C.S.; Lonsdale, W.M.; Fortune, J.; Maillet, J. (1998) Predicting weediness in a quarantine context. Comptes rendus 6eme symposium Mediterraneen EWRS, Montpellier, France, 13-15 Mai 1998, 33-40, and the Australian Quarantine Inspection Service website at http://www.aqis.gov.au/docs/plpolicy/weeds1.htm where a detailed description can be found. See also Walton, C.; Ellis, N.; Pheloung, P. (1998) A manual for using the Weed Risk Assessment system (WRA) to assess new plants. Australian Quarantine and Inspection Service.

The rainbow lorikeet (*Trichoglossus haematodus*) is a brightly coloured gregarious parrot, native to parts of Australia, Indonesia, New Guinea and east to New Caledonia. It became established around Auckland, New Zealand, following deliberate releases and supplementary feeding. The eradication campaign against the rainbow lorikeets in Auckland was contentious and public opinion was divided. Many people could see no harm in having an attractive addition to the local avifauna, but many others considered the potential risks to native birds unacceptable.

The following has been edited from http://www.doc.govt.nz/cons/pests/lorikeet.htm, the **New Zealand Department of Conservation** Fact Sheet: "Rainbow lorikeet feed primarily on pollen, nectar and fruits, but will feed on grains. They are prolific, rearing as many as three successive broods per pair in a single season. Australian horticulturists regard them as a significant pest, and in some states they are actively controlled. In Darwin 80-90% of some tropical fruit crops are lost to rainbow lorikeets. They could have a significant impact on New Zealand's horticulture industry.

Australian evidence, supported by reports from the people in Auckland, is that these birds are generally aggressive towards and dominate all other birds trying to use the same food source. Several NZ native species utilize the same food and nesting habitats as lorikeets. Native honeyeaters such as the tui, bellbird and hihi (stitchbird) use the same food sources, and can readily be displaced by a flock of lorikeets. The native stitchbird, kaka and kakariki (red-crowned and yellow-crowned parakeet) are also cavity nesters, so there will obviously be some competition for nest sites. Many of these threatened bird species are doing quite well on predator free islands in the Hauraki Gulf, well within flying distance of the lorikeets' release site. Lorikeets have been recorded travelling over twenty kilometres to Australian offshore islands and thus pose a significant threat to species whose survival is only possible on Hauraki Gulf island sanctuaries, which have been cleared of predators. The work of the Department of Conservation and thousands of volunteers over many years has been placed in jeopardy."

While the following has been edited from http://www.rainbow.org.nz/, the web site of the **Rainbow Trust**, founded by a group of residents who are convinced that the rainbow lorikeet has at least as much right to live in New Zealand as any of its critics: "A new compilation of evidence from a number of sources shows that the claims made by the Department of Conservation in various publications against the rainbow lorikeet are either exaggerated or incorrect. The evidence suggests that the rainbow lorikeet is anatomically not adapted to live and breed in the NZ bush, that it poses no threat to native birds by competition for food or nest sites, that the bird poses no threat to the horticultural industry, and that by publishing false information on this matter the Department of Conservation has misinformed and misled the Prime Minister, the Minister of Conservation and the public of New Zealand. The Minister of Conservation is requested to direct his Department to withdraw all misleading information and to correct the misinformation already broadcast about the rainbow lorikeet, and the inappropriate classification of the rainbow lorikeet as an unwanted organism under the Biosecurity Act should be withdrawn. The capture programme of the rainbow lorikeet would be an unnecessary waste of taxpayer's money and departmental resources."

There are now very few rainbow lorikeets left in "the wild', following Department of Conservation trapping, and probable recapture by the people that released them in the first place. Should they be released again, there could be fines of up to NZ$10,000 and/or a prison term for an offender.

Editors Note: We find that the views of the Department of Conservation are supported by published scientific information. We can find no published data to support the views of the Rainbow Trust.

CASE STUDY 3.22 Siberian Timber Imports: Analysis of a Potentially High-Risk Pathway

Siberia has almost half the world's softwood timber supply. In the late 1980s a few USA timber brokers and lumber companies, short on domestic supplies, wanted to bring in raw logs from the Russian Far East to West Coast USA sawmills. This could have created a pathway for non-indigenous forest pests that would be pre-adapted to many North American climate zones and tree communities. In the past 100 years raw wood or nursery stock imports have provided entry for a number of devastating pathogens into the USA, including chestnut blight (*Cryphonectria parasitica*), Dutch elm disease (*Ceratocystis ulmi*), and white pine blister rust (*Cronartium ribicola*).

In response to concerns raised by the scientific community about the risks of introducing exotic pests on logs from Siberia, the Animal and Plant Health Inspection Service (APHIS) imposed a temporary ban on Russian log imports in 1990 until a detailed risk assessment could be completed. A joint U.S. Forest Service/APHIS Task Force was convened and worked for almost a year on a detailed risk assessment focusing on larch (*Larix* spp.) from Siberia and the Russian Far East. The project involved 80 forest pathologists, entomologists, economists, and ecologists from federal and state agencies and universities and cost approximately US$500,000. The assessment identified many insects, nematodes, and fungi that would be potential pests if introduced into North America. The potential consequences of introduction were examined by considering the possible economic and ecological impacts should selected pests successfully invade north-west USA forests. For example, an estimate of cumulative potential economic losses from the Asian gypsy moth (*Lymantria dispar*) and the nun moth (*L. monacha*) between 1990 and 2004 is in the range of US $35 billion to $58 billion (net present value in 1991 dollars). The report concluded that: "measures must be implemented to mitigate the risk of pest introduction and establishment".

A companion report prepared by APHIS evaluated the possible treatments to mitigate the risk of importing exotic pests. This review identified many gaps in the scientific data on the subject and suggested that heat treatment appeared to be the best control option. The assessment concluded: "if technical efficacy issues can be resolved, APHIS will work with the timber industry to develop operationally feasible import procedures."

Ultimately, APHIS put the burden back on the importers to propose new pest treatment methods and protocols that "evidenced complete effectiveness" in mitigating risk. To date the industry has identified no feasible, cost-effective procedures that APHIS has deemed completely effective; thus, unprocessed logs from Siberia have been denied entry to the USA. While costly, the risk analysis was successful in preventing the potential introduction of several serious pests.

Adapted from U.S. Congress, Office of Technology Assessment, Harmful Non-Indigenous Species in the United States, OTA-F-565 (Washington, DC: U.S. Government Printing Office, September 1993), available at http://www.wws.princeton.edu/~ota/disk1/1993/9325. *See also "Pest Risk Assessment on the Importation of Larch From Siberia and the Soviet Far East, USDA Forest Service Misc. Pub. No. 1495, 1991" and "An Efficacy Review of Control Measures for Potential Pests of Imported Soviet Timber, USDA Animal and Plant Health Inspection Service, Misc. Pub. No. 1496, 1991."*

CASE STUDY 3.23 Invasiveness Cannot Be Reliably Predicted

Historically, invasion biologists have sought lists of species' traits that are likely to make an introduced species invasive – this is the holy grail of invasion biology, to be able to predict which introduced species will become problems and which will remain innocuous. From the 1950s through to the 1970s, efforts were based on lists of traits conducing to weediness or invasiveness, with such traits as high number of seeds seen as crucial. However, in the 1980s and 1990s, this approach was largely abandoned because it didn't work well. There were too many false positives and false negatives, plants that should have been weeds but were not, and plants that shouldn't have been weeds but became highly invasive. For example, among groups of congeneric species with very similar traits, often one is highly weedy and the others are not, e.g. *Eichhornia crassipes* (see case studies on water hyacinth, 5.1, 5.20, and 5.30). During the current explosion of interest in biological invasions, this idea has re-emerged, and recent studies have looked at traits for pines and for woody plants. In the case of pines, seed size almost completely determines whether a pine will become invasive.

Substantial efforts along these lines could constitute a step backwards, and may also lead to a false sense of security among managers and policymakers. In fact, we generally cannot predict very well which species will become invasive. This is not to say that we cannot do anything at the species level. The best predictor of which species will become problematic is whether or not a species has proven to be invasive elsewhere, especially under similar (climatic and geographic) conditions and in related ecosystems.

The difficulty of predicting invasiveness from species traits is exemplified by risk assessment as applied to individual species. The methods are still primitive, based on chemical models that do not account for such features of living organisms as evolution and autonomous dispersal, and estimates of risk are largely based on guesstimates by experts. There is a great need for research on risk assessment procedures for non-indigenous species.

The take-home message from this consideration of the invasion biology literature is that, alas, there are not many shortcuts in predicting which invasions will be problematic. There is no substitute to intensive biological research on the species in natural and invaded locations and the target community.

A further implication of the difficulty of such prediction is that black lists alone are unlikely to be an adequate tool. There are simply too many species that won't ever get on black lists that will nonetheless become invasive. So, although white lists are therefore likely to be an important approach, the requirements for white list entry must be very stringent, and it must be realized that, even with stringent requirements, there will be mistakes (cf. Section 3.3).

Edited from: Simberloff, D. "The ecology and evolution of invasive nonindigenous species", a paper presented at the Global Invasive Species Programme Workshop on Management And Early Warning Systems, Kuala Lumpur, Malaysia, 22-27 March 1999.

CASE STUDY 3.24   GISP Global Database / Early Warning Component

The GISP global database contains information on species, their taxonomy and ecology, their native and invaded distributions (including both habitat and location), impacts, contacts and references which can provide further information, plus reports on management methods. The opportunity to predict potential new invasions by matching habitat types with invaded range has been incorporated in the development of the GISP global database to contribute to prevention and early warning. In the future, it should be possible to add factors such as climatic suitability and pathways used, to further improve predictive capability and early warning potential.

The database is:

➤ Searchable (including by geographic zone, species, and generic variable e.g. "vine", "rat", contacts ...) and has a predictive component (by habitat match with invaded range).

➤ Accessible to low-tech users (e.g. "user-friendly", "browsable" readable information), as well as quick and reliable. Hardcopy versions.

➤ Satisfying to high-tech users (detailed data can be selected and retrieved to form specialised reports etc).

➤ Designed so that additions can be made in future (e.g. it will be able to generate an "alert list" of recently-introduced invasive species that are spreading rapidly across the region).

Future developments include a network of databases on IAS, a contribution to a thematic Clearing House Mechanism, dissemination and local adaptation of the global invasive species database, and improved predictive and early warning functions.

Prepared by Mick Clout, IUCN SSC, University of Auckland, New Zealand, m.clout@auckland.ac.nz, http://www.issg.org/database

Chapter 4

EARLY DETECTION

Summary

Early detection of non-indigenous species should be based on a system of regular surveys to find newly established species. However, not all species will become established, and only a small percentage of those that do will become invasive, presenting threats to biodiversity and the economy. Thus, some surveys will need to focus on specific target species known to be invasive under similar conditions or species that have been successfully eradicated before. Methods to detect species differ between taxonomic groups, and their success depends largely on taxonomic difficulties and how conspicuous species are. Sampling techniques are discussed for the major taxonomic groups. In addition, site-specific surveys looking for alien species in general can be carried out. They should be targeted at key sites, e.g. areas of high conservation value, within the range of highly endangered species, and at high-risk entry points such as airports and harbours. The drawback of these general surveys is that only well-trained staff will be able to identify non-indigenous species in many taxonomic groups.

Staff responsible for the surveys needs to be trained. Public education should focus on groups using or acquainted with the natural environment, such as farmers, tour operators, and the concerned public. This education campaign can be based on media promotion, displays, and personal interactions. The training of survey staff must include development of taxonomic knowledge, use of databases and identification services, and survey methods for the different groups. The training could be either in-country, with or without overseas experts, or in courses held abroad.

A crucial part of early detection is a contingency plan, which determines the action to be taken when an alien species is been found. Given the diversity of potential new incursions, an initial plan will be rather general. It should summarize the stakeholders and experts who need to be contacted for a more detailed action plan. Contingency plans targeted at specific high-risk species can be very efficient, with an exact schedule for what to do. For a contingency plan to work, the equipment needed must be in perfect condition and at the designated place. The relevant government departments responsible for bioinvasions should make contingency funding available for emergency eradication or control.

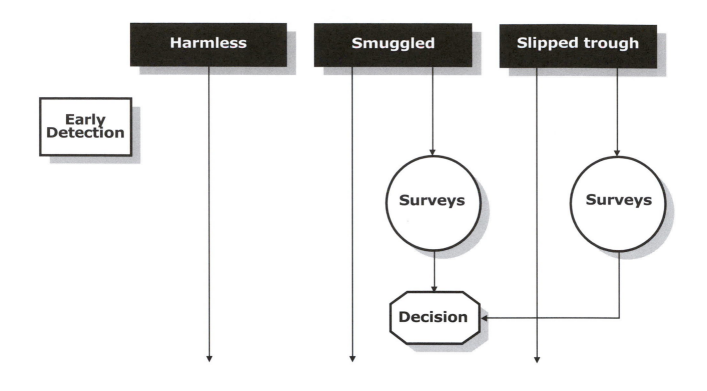

Figure 4.1 Species which are present in the country in spite of prevention measures (cf. Figure 3.1) need to be detected early in their establishment phase using appropriate survey techniques, in order to decide what to do before they become invasive. Established alien species will then belong to one of the following categories: intentionally introduced white list species, undetected species, and species detected in surveys (see figure in the Toolkit Summary for the full flowchart).

Introduction

Once an alien species is present in a new country, there will be a brief period when its chances of establishment will hang in the balance. However, the longer it goes undetected at this stage, the less opportunity there will be to intervene, the fewer options will remain for its control or eradication, and the more expensive any intervention will become. For example, eradication (discussed in Section 5.3.1) will rapidly cease to be an option the longer an alien is left to reproduce and disperse. Not all alien species will necessarily become invasive, so species known to be invasive elsewhere, especially those spreading within a region, should be priorities for early detection. The possibility of early eradication or getting a new colonizer under effective early control makes investment in early detection worthwhile.

4.1   Surveys

If new invasive species are to be detected at an early stage, then surveys are needed. Surveys for early detection should be carefully designed and targeted to answer specific questions as economically as possible. They are not necessarily intended to collect scientific data, but will usually be designed to give a yes or no answer. Beware of getting locked into a self-sustaining programme of doing surveys for the sake of surveys.

Some invasive species are easily seen while others are cryptic and require special efforts to locate or identify them, particularly when they are in low numbers. Visitors knowledgeable about invasive species in other areas may be the first to draw attention to a new invasive (Case Study 4.1 "First Detection of European Green Crab in Washington State"), but waiting for some person to happen upon and report a new invader often means that the invader will be well established by the time relevant authorities become aware of it. Surveys by experts should be made for certain groups of pests to enable a rapid response before the species becomes well established.

For a detailed treatment, see the one produced by the International Plant Protection Convention (1997. Guidelines for Surveillance. International Standards For Phytosanitary Measures, 6. Secretariat of the International Plant Protection Convention, Food and Agriculture Organization of the United Nations, Rome, 15 pp. Also available at http://www.fao.org/ag/agp/agpp/pq/default.htm under International Standards for Phytosanitary Measures).

Three types of surveys can be considered: general surveys, site specific surveys and species specific surveys. Depending upon the purpose, these categories may merge or overlap, e.g. species specific surveys may be carried out in a site specific way.

4.1.1   General surveys

For large or conspicuous animals and plants this is a "looking survey". While doing other work, staff should be vigilant and continually aware of possible signs of new invaders. Encourage public reporting of new sightings (Case Study 4.2 "Early Detection and Eradication of White-Spotted Tussock Moth in New Zealand"). The conservation organization can then identify the species and report back to the member of the public to maintain good public relations. Encourage interest groups, such as botanical societies, to undertake specific searches for new species.

4.1.2   Site specific surveys

These could be characterized as general surveys targeted at key sites, e.g. high value biodiversity areas and areas near high-risk entry points. Entry points are considered in more detail under Chapter 3 "Prevention". For terrestrial systems, high-risk entry points include airports, seaports, and container or freight unpacking areas, whereas harbours are the main entry points for marine species. High value areas may be entire reserves or small and valuable habitats where you will either want to try to exclude new arrivals or document environmental impacts of new arrivals that cannot be controlled. River corridors may be entry points to reserves. This documentation can be valuable for strengthening preventative methods. The survey needs to extend beyond the entry point, depending on the habitat, geography, tracks and roads around the entry point. These methods of survey are somewhat generalist since we don't necessarily know what we are looking for but we do want to find it if present.

Important sites for land animals. Search for vertebrate signs, such as tracks, droppings and feeding damage. Know your fauna and look for new species. Know who the local experts and contact people are. If you find or suspect a new species, record it carefully, report it and ensure that it is identified rapidly.

Plant communities at high-risk. The best method is to use an experienced botanist who knows the botany of the area. This person should be able to readily identify a new arrival (Case Study 4.3 "Early Warning Systems for Plants in New Zealand). For people with less botanical knowledge the provision of identification aids is essential. These aids in the form of books, field guides and posters need to target known prior invaders, invasive species which are present in neighbouring countries, easily transported species and invaders of similar bio-climatic zones. It may be necessary to rank or group these species to assist staff to learn their identification.

Marine environment. In view of the lack of success in controlling aquatic invasives once established, early detection is likely to be less of a priority. As indicated elsewhere in this guide, prevention is the most effective strategy in this situation. Nevertheless eradication can be considered if a new invasive can be spotted and recognized early enough (Case Study 4.4 "The First Eradication of an Established Introduced Marine Invader").

The only general surveys we know of for exotic marine species are irregular, ad hoc, poorly-funded or unfunded efforts, when a team of marine taxonomists (typically with many of them volunteering their time) have got together to examine a series of stations in a designated region over a short period of time (about a week), specifically looking for exotic organisms. This would generally be focused on dock fouling because it can be quickly and effectively sampled by a team of people without regard to the tide level, so that several stations can be sampled quickly. In the USA four such surveys have been carried out in San Francisco Bay, California and two in Washington. See also Case Study 5.23 "Eradication of the Black Striped Mussel in Northern Territory, Australia" in which regular monitoring that would detect new alien species is carried out in a harbour, in the context of a recent eradication.

4.1.3 Species specific surveys

Where specific threats are identified and prioritised, it will be appropriate to make regular surveys that are carefully planned using specific methods in potential habitats of possible invaders. The methods are very specific and will need to be designed, adapted or developed for each situation. Frequency and timing of surveys is important. The potential range of newly arrived invaders needs to be considered along with the climate of the region. In equable climates new invaders may be difficult to detect at all times of the year so more frequent or more diligent survey will be needed. In highly seasonal areas new invaders are less likely to establish in winter, while plants may not be identifiable without their foliage, so annual surveys may be sufficient.

Plants. Survey methods for specific plant invaders will depend on how easy it is to recognize the target. Sometimes it will be very easy, but if there are similar non-invasive and/or indigenous species present, then field guides, illustrations and training may all be necessary (Case Studies 4.5 "Detection of Chromolaena Weed in Australia" and 4.6 "Public Awareness and Early Detection of *Miconia calvescens* in French Polynesia").

Mammals. The best methods differ for each animal group. Large invasive vertebrates, such as goats and cattle, are reasonably visible, leave noticeable and distinctive sign, have a low reproductive rate and can take some time to have a significant impact on an ecosystem. An annual or biennial survey by a knowledgeable observer walking and looking should detect their presence.

Smaller invasive vertebrates, such as rodents and feral cats, are much less visible, leave sign that can be hard to find, have a moderate reproductive rate and can quickly spread throughout an ecosystem and have a significant impact on that ecosystem. Some, if not most, small vertebrates are difficult to detect when they are in low numbers. Surveys need to be species-specific, seasonally timed, habitat selective and quite intensive (Case Study 4.7 "Early Detection of Rats on Tiritiri Matangi").

There are occasions when species that are very similar to an existing native species will invade or have a potential to invade. An example is when the house mouse (*Mus musculus*) invades an island that is already populated with native mice, such as *Peromyscus* of America. Methods to detect such invasions and identification of new invaders need to be done with great care in this situation.

Insects. There are survey methods that will catch a broad range of insects and other invertebrates. These are unlikely to be useful unless the insect being surveyed is conspicuous or specialists are available to monitor the trap catches. It is more appropriate to design survey methods to suit the insect being surveyed, based on specific behaviours or characteristics of the invader (Case Study 4.8 "Early Detection Plan for Hibiscus Mealybug in the Bahamas"). Sometimes, very specific and effective trapping methods are available, e.g. pheromone traps or targeted lure traps, and these can be used to locate new arrivals effectively.

However, rather than try to provide comprehensive species-specific advice here (a task beyond the scope of this guide), managers with this problem are advised to collaborate with local entomologists to formulate appropriate strategies.

Reptiles, e.g. lizards and snakes. Specific survey methods may need to be developed for early detection of these species. Trapping using rodents as bait in double-compartment traps has been used on Guam to survey for the brown tree snake, and is being used for early detection on adjacent islands (Saipan and Rota) that are at risk of invasion. General survey and a high level of public awareness are important.

Freshwater fish and invertebrates. Occasionally biologists conducting routine sampling are the first to encounter a new freshwater organism, however, most often it is the general public who catches or finds something they can't identify and reports it. Anglers can be very useful in detecting fish introductions.

Fish and invertebrate sampling techniques vary depending on the habitat, depth of the water, and species sought. Possible sampling methods for fish include gill nets, trawls, seine nets, rotenone, angling and electro-shocking. Invertebrates are more likely to escape notice because many of them are small. Sampling techniques range from ponar grabs for benthic organisms to plankton tows for planktonic organisms.

Marine fish and invertebrates. Again, we are not aware of any procedures in place for specific surveys to detect species of alien marine fish. Known invasive alien invertebrates can be monitored as they spread (Case Studies 4.1 "First Detection of European Green Crab in Washington State" and 5.23 "Eradication of the Black Striped Mussel in Northern Territory, Australia").

Pathogens. Species specific surveys have been organized for diseases of agricultural importance, e.g. rubber blight, witches broom of cacao, coffee leaf rust, as they spread around the world, and these could provide models for monitoring for early detection of diseases of environmental importance.

It should not be forgotten that some diseases are vectored by insects or other animals. A country can have the disease and not the vector or vice versa. In the former case, the disease would only become a problem if the vector arrived, and so it is the vector that would need to be monitored. Thus citrus tristeza virus has been present in Latin America and the Caribbean for a long time, but only came to prominence recently when its aphid vector also colonized the region (Case Study 4.9 "Spread of the Aphid Vector of Citrus Tristeza Virus").

Ecosystem Surveys. It may be suggested that survey of an ecosystem will detect the presence of invaders through observation of diminishing prey species or skewed age classes. For example if rats are present some birds will produce no young, or if snakes are present small native mammals or small birds may diminish in number. While such changes are likely, they will not occur until the invader is established well beyond the "early detection" stage. Hence ecosystem surveys are not recommended for early detection of invasive species.

4.1.4   Data collection and storage

For all these surveys, it is important to keep a good record of the species found, both native and introduced, and the action taken. In the case of most groups of invasive species, voucher specimens should be collected and preserved. When local knowledge is not adequate to make an authoritative identification, material should be sent for specialist identification. Local and regional museums are a good starting point for advice on identification of invasives, but there are also specialized international services available. BioNET International is a global network for capacity building in taxonomy, and contact with your local network or LOOP may help identify regional expertise (http://www.bionet-intl.org/). The Expert Centre for Taxonomic Identification (ETI) maintains an internet database of taxonomists (http://www.eti.uva.nl/database/WTD.html). The Global Taxonomy Initiative recently started under the Convention on Biological Diversity will also be a valuable resource in this area in future. Establish and keep up-to-date a contact list for your country or region. This needs to include the names of both institutions and people, what types of invasive species they might be able to identify and the methods that should be used for the specimen collection.

Collect the data in a standard format and store it in a national database. The data fields of the Island Invasive Species Database can be used as a format for data collection. In this way both positive and negative species locations are recorded (See also Case Study 4.13 "Building a Knowledge Base for Rapid Response Action").

4.2   Developing a corps of experts/trainers

4.2.1   Who to train

There are two main groups of people to consider in order to develop a national capability in early warning:

Group 1: those who are tasked nationally with the surveying and scouting that is needed. This will vary from country to country, but is likely to be based upon National Parks Officers and Conservation Managers or the national equivalent. Thus in New Zealand, biosecurity officers and conservation officers would be the targets, in Malaysia surveillance (at least for agricultural exotics) comes under the Department of Agriculture. In a country of the size and complexity of the USA, the situation is less straightforward. Several agencies or organizations have responsibility for, or interest in, the problem, including Fish & Wildlife, USDA-ARS, USDA-APHIS, and USDA Forestry Service. A wide range of expertise is needed, including entomologists, pathologists, foresters, botanists, freshwater and marine scientists, etc., which will be spread across several or many groups and organizations.

In many countries, there may be no mandated service, and this is an area that would need to be addressed in a national strategy in consultation between environmental and agricultural Ministries (since they will already handle this problem for agricultural pests).

Group 2: others who could notice new aliens in the course of their activities. This is a much larger group, and could include:

➤ the concerned public, especially those interested in natural history,

➤ farmers,

➤ gardeners and landscape managers

➤ forestry field staff,

➤ fishermen (subsistence, recreational and commercial),

➤ ecologists, natural history clubs and environmental groups,

➤ land surveyors,

➤ educators,

➤ dive instructors and tour boat operators,

➤ tourist operators,

➤ hiking clubs and climbers,

➤ photographers,

in fact, almost anyone who spends time in the natural environment and has time and opportunity to observe the flora and fauna around them. These people will need awareness raising activities as much as training, and in general developing capability will have to be handled by those trained under Group 1 (Case Study 4.10 "Community Monitoring of Introduced Marine Pests in Australia").

There are many ways to develop capability and awareness in Group 2, which might include activities such as:

➤ media promotion,

➤ availability of field guides,

➤ personal interactions,

➤ displays at nature reserves, museums, etc.,

➤ field trips to sites where there are invasive species,

➤ making fact-sheets available, both as hard copy and on the internet,

➤ preparation of school materials, posters etc. (for example, see http://www.aphis.usda.gov/oa/alb/albposter.pdf for a poster on the Asian long horned beetle in North America, and Case Study 4.11 "Public Awareness Poster for Cypress Aphid").

4.2.2 Training needs

The needs for Group 1 are considered here. This group of professionals will benefit from in-country training, probably using regional and overseas experts. The main objective of this training would be to generate the capacity to identify native and foreign organisms. Areas to cover include:

➤ general training to increase knowledge of native species and hence enhance identification of new species,

➤ training to identify aliens on a black list;

➤ training in use of databases, keys, manuals and other identification sources, to identify aliens known to be invasive elsewhere;

➤ recognition of the presence of new species;

➤ how to collect, label and preserve suspected invasives for identification;

➤ how to get things identified; and

➤ the concept of cryptic species and what to do about them.

In support of this training, specialists will be needed to prepare data sheets on identified high-risk invasives. The GISP invasive species database (Case Study 3.24 "GISP Global Database / Early Warning Component") and the ISSG would be important sources of materials for this. In addition various groups have already put data sheets on the internet (Box 2.1 "Some Internet-Based Databases and Documents on Invasive Alien species"), and some of these have been used as information sources for the Case Studies provided in this toolkit. Invasive species also of concern to agriculture and forestry will be covered by information sources intended for this sector, e.g. the CABI Crop Pest Compendium (Box 5.3).

4.2.3 Where to train

For training staff from Group 1, it would be desirable to hold the training either within the country, or in the case of under-resourced countries within the relevant region. Thus, for Pacific Small Island Developing States (SIDS) one might choose Hawaii as a place that has plenty of likely invasives, and the capacity to assist with training. Training as far as possible must be site or region specific.

Those in Group 2 would be "trained" in-country by Group 1 with the use of specifically prepared materials, and the support of the media.

4.2.4 Who will do the training

Country or region specific courses will usually need external inputs from resource people. These will most likely come from neighbouring developed countries, international organizations, universities etc., and need to be government or donor funded.

Group 1 will train and expose Group 2 - providing the individuals in Group 1 have or can develop adequate teaching skills.

4.2.5 Staff retention

In the case of SIDS in particular with limited human resources, the retention of staff who have been trained is a significant factor. There are of course many locally varying factors and policies that will affect this, and there is little advice that can be offered in this forum except:

➤ If possible train more staff than are immediately necessary, and

➤ once they have gained experience and their knowledge and skills have been assessed, then the trainees should be able to train others in turn.

4.3 Contingency plans and funding

A contingency plan is usually a carefully considered outline of the action that should be taken when a new invasive species is found or an invasion is suspected. Given the diversity of potential invasive alien species, and the variety of options for strategy and control methods for different species, on pragmatic grounds, plans will initially have to be either very broad-brush – perhaps identifying general principles, responsibilities and the likely stakeholders who would need to convene to draw up a detailed action plan in response to a specific event - or targeted towards specific potential invasive species or groups identified as high risk. Over time, more specific components for different groups or species can be added to the overall plan to provide a detailed contingency plan for more general use. Of equal importance to the contingency plan is the involvement and commitment of all the people involved in caring for the area at risk. They must all understand the plan and, to the extent that it includes prevention and early detection, put parts of it into effect every day.

Specific plans can be very simple, for example, the USDA-APHIS plan to manage hibiscus mealybug (Case Studies 4.8 "Early Detection Plan for Hibiscus Mealybug in the Bahamas" and 5.11 "Colonization Rate of Hibiscus Mealybug in the Caribbean") when it arrived in mainland USA was very straightforward: biological control, based on the experience gained with, and documentation prepared for, the Caribbean would be implemented. As a result, when hibiscus mealybug was first reported from California in 1999, the first biological control agents were introduced three weeks later.

The contingency plan may be just a simple paper document that staff, selected volunteers and nominated other organizations have written, are aware of, and will act on in a contingency situation (Case Study 4.12 "What Goes into a Contingency Plan?"). Or the plan may expand to include comprehensive kits of tools that are stored in a "ready to use" condition at appropriate locations, e.g. USDA-APHIS's cultures of hibiscus mealybug parasitoids, or perhaps a supply of rat poison and bait for a rapid response eradication programme. The equipment needed for contingency action must be maintained in perfect working order and stored where the plan says it should be.

To prepare the plan, possible contingency situations and actions that could be taken need to be considered and possible actions discussed and agreed to by all parties. Examples of action to be taken are:

Plant examples. A suspected new invasive is found. It is just one plant. The finder is a botanist and knows that it is a new invader. The contingency plan says the plant should be pulled up and put in a secure container on the site to avoid seeds and bits being dropped. Then it should be taken to the quarantine station and burned. The site it came from should be carefully marked and checked every six months for the next two years.

A suspected new invasive is found. It is a small patch of plants. The finder is a conservation officer who is not a botanist and is not sure whether it is a new invasive or a very rare native plant. The contingency plan says a small piece should be collected and taken to a botanist. Every endeavour should be made to identify the plant within three days of it being found. If it is an invasive the contingency plan says all flowers and seeds should be removed and put in a secure container on the site to avoid seeds and bits being dropped. Then these should be taken to the quarantine station and burned. The site it came from should be carefully marked and action to manage the plant considered by following standard assessment procedures.

Mammal example. A ship is wrecked on the coast of a pristine island. The contingency plan requires that a team of people take the pre-prepared rodent contingency kit to the island and place rodent baits and traps around the wreck site according to carefully prescribed instructions. They then check the wreck for rat sign and consider how long to continue to trap and bait the area.

Reptile example. A snake is reported from an island where snakes do not naturally exist. The contingency plan requires that a team of people take the snake contingency kit to the island and hunt for the snake. At the same time they should involve the person who reported the sighting in a positive and friendly way while questioning the person for details. The contingency plan may then require that a quarantine officer with a trained snake dog be brought in for further searching.

Invertebrate example. When a potential invertebrate invasive is prioritised, contingency plans can be prepared to respond to their arrival (Case Study 4.13 "Building a Knowledge Base for Rapid Response Action"). There are many examples from agriculture which can be used for guidance, e.g. biological control of hibiscus mealybug mentioned above. Many more scenarios for different groups of potential invasive alien species would need to be considered in the formulation of a contingency plan.

4.3.1 Costs of contingency actions

Every conservation organization with responsibility for pristine islands and reserves where new invaders are likely should financially support the creation of a contingency plan and assembly of contingency kits. There are probably no organizations that can have money set aside for possible contingency action. The contingency plan needs to detail how the costs of contingency operations should be met. For example:

➤ If the action taken involves staff time and costs of less than a notional amount (say $50) then the staff should just get on with the job.

➤ If the action required involves the next level up of time and expenditure (say $500) then the action will still proceed and the manager will need to decide what other work will not be done to compensate.

➤ More expensive options will need authorization, and since mobilization of these resources usually will take at least a few days, then time should be available to generate approval of implementation of the prepared contingency plan.

The relevant government departments responsible for responding to new invasions should consider setting in place mechanisms (laws, regulations, authorities and responsibilities) to make available the necessary funds for rapid deployment of resources to deal with emergency control operations. A far-sighted department would have these in place before an emergency happens. For example, the US Department of Agriculture has the authority, under law and regulation, to use any available funds for emergency control or eradication.

CASE STUDY 4.1 First Detection of European Green Crab in Washington State

In 1998, researchers in Washington State, USA, invited Andrew Cohen, a marine ecologist with the San Francisco Estuary Institute in Richmond, California, to visit and survey a stand of non-native cord grass that was invading the shallow waters of Willapa Bay, in south-western Washington.

Just 30 minutes after donning his boots and wading into the water, Cohen stumbled upon the first evidence that an even less welcome invasive alien species had reached Washington. He had found the moulted shell of a male European green crab (*Carcinus maenas*), which from his experience in California he was immediately able to recognize.

There had been an enormous - almost unprecedented - amount of publicity about the expected arrival of the green crab in Washington State, and the anticipated harm that it would do to the shellfish industry and to coastal ecosystems. A great many local people were thereby alerted and encouraged to keep an eye out for green crabs. Yet, it was an outside expert on invasions, familiar with that particular crab, on a visit for a very brief inspection of the coast, who found the first green crab in the state.

Prepared from information provided by Andrew Cohen, San Francisco Estuary Institute, 1325 South 46th Street, Richmond, CA 94804 USA: http://www.sfei.org/invasions.html.

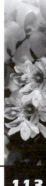

CASE STUDY 4.2   Early Detection and Eradication of White-Spotted Tussock Moth in New Zealand

The very distinctive caterpillar of White-Spotted Tussock Moth (*Orgyia thyellina*) was first collected by a member of the public from a peach tree in suburban eastern Auckland in April 1996.

Native to Japan, Taiwan and Korea, this insect had adapted to conditions in this sub-tropical northern region of New Zealand following its accidental introduction one or two years before, and had the potential to cause severe damage to a wide range of trees and other plants.

Surveys by New Zealand's Forest Health Advisory Services team showed that the distribution of this new pest was limited to an area of about 100 ha. The New Zealand Ministry of Forestry led a multi-agency contingency response process based on an eradication programme using Foray 48B (*Bacillus thuringiensis var. kurstaki*) as a one-chance control option. Foray 48B was applied using ground and aerial techniques, initially over 4000 ha, but this area was progressively reduced so that final applications were contained to only 300 ha.

Not surprisingly, overwintering egg masses failed to display a natural Northern Hemisphere synchronous hatch pattern, leading to a spraying programme commencing in October 1996 and extending into early March 1997. In all, 23 aerial and associated ground-spraying treatments were applied to the infected and buffer areas. At all times, the eradication operations were fully supported by a team of relevant research and technical experts working alongside operational and media specialists. Monitoring of spray efficacy was achieved using a variety of methods. For example, female moths were confined in sealed traps at secure locations throughout the region, in order to attract males. Six males were trapped in April 1997, but no live stages of O. *thyellina* have been intercepted in the field since then.

A parallel international initiative instituted by the Ministry of Forestry led by mid-1997 to the isolation and synthesis of the pheromone used by female moths to attract males. This enabled the Ministry to set out an array of 7,500 pheromone baited traps during the summer of 1997-98. No moths were found, and the project was wound up in July 1998. A sentinel pheromone trap array was maintained during the 1998-99 summer again with no O. *thyellina* captures, and the programme is considered to have eradicated the moth in New Zealand.

The NZ$ 12 million spent on the programme is considered justified, on the basis of the undoubted impact which the moth would have had on the urban forest environment, horticulture and, to a lesser extent, the exotic and indigenous forests of New Zealand.

Ross Morgan, National Manager, Forest Health Forest Health Advisory Services, PO Box 6262 Rotorua, New Zealand. E-mail: MorganR@forestry.govt.nz

CASE STUDY 4.3 Early Warning Systems for Plants in New Zealand

Auckland Region Council Biosecurity Unit staff carry out an annual check of the facilities of 274 plant growers and retailers in the Auckland Region. They search for and confiscate any illegal plants, check out new potential invasives and look for untidy work practices that may result in unnecessary spread of plant species.

Auckland Region Council Biosecurity Unit staff annually check a new 5% of the land area of the Auckland Region searching for weed infestations. This work began after establishing higher risk priorities as the lands of peri-urban areas and beside high value conservation reserves. This 5% may seem to be a small area but when dense housing, clear farmland and dense forest is considered as low risk the high risk areas will be closely inspected every three or four years.

Prepared by Dick Veitch, Papakura, New Zealand.

CASE STUDY 4.4 The First Eradication of an Established Introduced Marine Invader

An unknown species of sabellid polychaet annelid inadvertently arrived in California in a shipment of abalone from South Africa. This pest was initially contained in mariculture facilities. The worm causes shell deformation and slowed growth in the cultured abalone. Established intertidal populations were discovered near Cayucos, California, in 1996.

An eradication programme based on the "epidemiological theory of threshold of transmission" was implemented and defined as when the density of transmissive stages and the density of highly susceptible hosts are reduced below the replacement transmission rate, successive generations of the pest will die out.

The eradication programme included (1) prevention of further release of adult worms from the facility; (2) reduction of the adult pest population; and (3) reduction of the most susceptible native host population. The three-way approach targets the pest but also the host that is required for continuance of the established population. In April 1998, surveys found that new infestations had been eliminated. This potentially successful eradication programme suggests the need for (1) early detection; (2) co-operation between commercial interests, regulatory agencies, and pest control scientists; (3) rapid response; (4) development of control strategy with a theoretical basis; (5) persistent efforts beyond the point where the situation has merely improved; and (6) monitoring of eradication efficacy through use of sentinel habitat experiments.

Abstracted from: Culver, C. S. & A.M. Kuris (1999) The Sabellid Pest of Abalone: The First Eradication of an Established Introduced Marine Bioinvader? In Marine Bioinvasions, Proceedings of the First National Conference, J. Pederson (Ed.), January 24-27, 1999, Massachusetts Insitute of Technology, MIT, Cambridge, pp. 100-101.

CASE STUDY 4.5 Detection of Chromolaena Weed in Australia

'Chromolaena' or Siam weed (*Chromolaena odorata*) is considered to be one of the world's worst weeds and has the potential to spread across northern Australia and down the eastern coastline. If established in Australia, Siam weed will seriously degrade large areas of the wet/dry tropic savannah grasslands, and conservation areas. Agricultural and horticultural production, sugar cane and forestry plantations are also at risk.

Several small infestations of chromolaena were discovered in 1994 in the Tully River district of far northern Queensland. On 15 July 1994, several flowering plants were noticed growing along a roadside near the coastal village of Bingil Bay. Samples were collected and verified against herbarium specimens held at the Department of Primary Industries.

The likelihood of chromolaena's arrival in tropical Australia had been recognized and predicted in recent years. Personnel employed on the Northern Australia Quarantine Strategy (NAQS), had looked for chromolaena in remote parts of northern Australia (e.g. the Torres Strait Islands) and in neighbouring Papua New Guinea. Familiarity with chromolaena from NAQS surveys in Papua New Guinea enabled the discoverer to recognize it and notify authorities immediately.

At the time of the discovery it was apparent that the plants observed were not the primary infestation, and must have spread from elsewhere in the district. An intensive survey programme conducted jointly by personnel from the Queensland Department of Lands and Queensland Department of Primary Industries has subsequently delineated the extent of the infestations. The primary infestation was discovered on pastoral lands in the vicinity of Echo Creek, a tributary of the Tully River, and various secondary infestations were also located.

Anecdotal reports from landholders along the Tully River suggest that chromolaena plants have been present on the riverbanks several kilometres downstream from the mouth of Echo Creek, for at least 7 years. This implies that the primary infestation in the upper reaches of Echo Creek is probably older than 10 years. Senescent plants with basal stem diameters of 7-10 cm have been located in this area. The infestation of chromolaena near Bingil Bay is estimated to be approximately five years old.

Time has obscured the clues as to the initial means of introduction of chromolaena, although contaminated pasture seed from overseas is the most likely source. Contaminated agricultural machinery or travellers returning from overseas are alternative sources.

Edited from: http://www.dnr.qld.gov.au/resourcenet/fact_sheets/pdf_files/pp49.pdf, the DNR Pest Facts web-page on Siam Weed and unpublished DNR reports.

CASE STUDY 4.6 Public Awareness and Early Detection of *Miconia calvescens* in French Polynesia

Since the recognition by local authorities (French Polynesian Government and French High Commission) of the severity of the invasion by *Miconia calvescens* on the islands of Tahiti and Moorea (French Polynesia) (cf. Case Study 2.6), an *M. calvescens* research and control programme started in 1988.

Three information and education posters ("Le Cancer Vert" in 1989, "Danger Miconia" in 1991 and "Halte au Miconia" in 1993) were published by the Department of Environment and widely distributed to all 35 high volcanic islands of French Polynesia susceptible to invasion. Each year, researchers displayed an information board on the *M. calvescens* programme during popular events in the town of Papeete, Tahiti ("Environmental Day" in June, " Agricultural Fair" in July, "Science Festival" in October).

Active manual and chemical control operations started in 1991 in the newly invaded island of Raiatea, where the Rural Development Service discovered small infected areas in 1989. By now, 6 annual campaigns have been organized in Raiatea with the help of hundred of schoolchildren, nature protection groups and the French Army. The 5-days campaigns were publicized in local newspapers, by radio and above all on a local TV channel (RFO 1 which is watched in all the inhabited islands of French Polynesia) during the television news in both French and Tahitian languages.

As a direct result, a small population of *M calvescens* was found and reported in 1995 by a pig hunter in a remote valley on the island of Tahaa, and local inhabitants noticed *M. calvescens* seedlings on the island of Huahine. In June 1997, during a botanical exploration in the Marquesas Islands conducted by the Research Department and the National Tropical Botanical Garden (Hawaii), a small population was discovered and destroyed on Nuku Hiva. Once again, an article was published in the local newspapers and a talk was made on local radio stations (including the Marquesan radio).

During the 4 days of the first Regional Conference on *Miconia* Control held in Papeete, Tahiti in August 1997, local TV, newspapers and radio have been again highly involved. As a result, more isolated plants were found and reported in the remote islands of Rurutu and Rapa (Austral archipelago) and Fatu Hiva (Marquesas archipelago) and immediately destroyed by the Department of Agriculture.

Prepared by Jean-Yves Meyer, Délégation à la Recherche, B.P. 20981 Papeete, Tahiti, French Polynesia. E-mail Jean-Yves.Meyer@sante.gov.pf

CASE STUDY 4.7 Early Detection of Rats on Tiritiri Matangi

Pacific rat (*Rattus exulans*) was eradicated from the 200ha island Tiritiri Matangi off New Zealand in 1993. The island is now carpeted with regenerating indigenous seedlings. The early detection method for possible re-invasion by rats is 100 rodent bait stations around the shoreline. These are checked monthly and the bait replaced every three months.

Prepared by Dick Veitch, Papakura, New Zealand. See
http://www.doc.govt.nz/cons/pests/fact51.htm#top and
http://www.doc.govt.nz/cons/offshr/off1.htm for more information on the rats.

See also Case Study 5.34 "Ecotourism as a Source of Funding to Control Invasive Species " for another example.

CASE STUDY 4.8 Early Detection Plan for Hibiscus Mealybug in the Bahamas

Hibiscus mealybug (*Maconellicoccus hirsutus*) is an Asian insect that was found in the Caribbean island of Grenada in the early 1990s. It has a wide host range and caused severe damage to ornamentals (notably hibiscus), agricultural crops (e.g. cacao, okra, mango, plums, sorrel, soursop), amenity trees (e.g. samaan), forestry trees (e.g. teak), and watershed trees (e.g. blue mahoe). It also started to spread and had reached the Virgin Islands when the Bahamas, in discussion with CAB International, put together the following plan for early detection:

➤ Monitor for new infestation at key high-risk entry points. On the basis of the current known distribution of hibiscus mealybug these are the air and seaports of Nassau and Freeport, and the seaports of Inagua and Exuma.

➤ Monitor in the vicinity of the dump used for cruise boat garbage.

➤ Trap plants such as hibiscus can be planted in the vicinity of the above areas if not already present and form the basis of a regular monitoring programme.

➤ A public awareness programme should be used to alert the public of the risk and implications of hibiscus mealybug coming to the Bahamas.

➤ Encourage the public to report symptoms of hibiscus mealybug infestation to the Ministry, perhaps through a dedicated hotline.

In a worst-case scenario, the Bahamas will learn about the presence of hibiscus mealybug in their country when one of their trading partners intercept hibiscus mealybug on Bahamian produce.

The initial reports of infestation will need to be checked by Ministry staff, and where it does indeed appear that the hibiscus mealybug may be present, this will need to be identified by a competent authority as the essential next step before putting into action prepared plans to address the problem.

Extracted from an unpublished report on a Ministry of Agriculture and Fisheries, Bahamas, and CAB International Workshop, July 1997.

CASE STUDY 4.9 Spread of the Aphid Vector of Citrus Tristeza Virus

Citrus Tristeza Virus (CTV) is a disease of citrus caused by a closterovirus, which resides in the phloem tissue. A common form is decline of scion varieties grafted onto sour orange rootstock. This can be very rapid, i.e. within weeks, when it is referred to as "quick decline". The problem can be so severe that in some countries sour orange has been abandoned as a rootstock. Even when the rootstock is tolerant, stem pitting due to tristeza can result in stunted trees with low vigour and small, worthless fruit.

The most efficient vector is the aphid *Toxoptera citricidus*. This Old World species has been in South America for some years and reached Central America in 1989 and has been spreading through Central America and the Caribbean since then. When newly arrived in an area it rapidly builds up large populations on flush growth and is then very conspicuous because of its black colour. Thus, it is very simple to spot by carrying out regular inspections.

Transmission of tristeza is non-persistent or stylet borne, which means that the aphids can only transmit the virus for 24-48 hours after acquiring the virus by feeding on an infected tree. Dispersing aphids that arrive in a new area are likely to be free of the virus. If the virus is already present at a low and uneconomic frequency as is normally the case, the aphids will rapidly acquire the virus and spread it to all trees.

The situation in Venezuela was especially severe. CTV was first reported in Mexican lime germplasm collections in Venezuela in 1960, but it was not a commercial problem at that time.   In 1976 *Toxoptera citricidus* was discovered in Venezuela for the first time, entering from both the south (Brazil) and west (Colombia). During the 1970s efforts were made to warn growers of the potential time bomb that CTV represented, but they were slow to react to the severity of the problem.

In 1980 the time bomb went off when the first serious CTV outbreak occurred. Twenty-four percent of trees sampled in 1980 by ELISA were seropositive. In 1981, 49% were seropositive, and this rose to 64% and 72% over the next two years. Over five million trees were lost by 1991. This crisis in the citrus industry provoked a belated major change to tolerant rootstocks. From having 99% sour orange in 1970, by 1992 only 10% of the rootstocks were sour orange.

Despite the change to tolerant rootstocks, CTV is still a major concern. Additionally, citrus blight (sudden decline) has occurred in high incidence on CTV tolerant rootstocks. Viroids and psorosis limit productivity of other CTV tolerant rootstocks and scions.

Edited from Lee, R.F.; Baker, P.S.; Rocha-Peña, M.A. (1994) The citrus tristeza virus (CTV). International Institute of Biological Control, Ascot, UK.

CASE STUDY 4.10   Community Monitoring of Introduced Marine Pests in Australia

Over 150 introduced marine species have now been discovered in Australian waters. Eight of these are recognized as major marine pests: *Asterias amurensis* (northern Pacific seastar), *Undaria pinnatifida* (Japanese seaweed, "wakame"), *Sabella spallanzanii* (giant fan worm), *Carcinus maenas* (European green crab), and four species of toxic dinoflagellates. However, there are at least 14 introduced species in total that are suspected of posing an environmental threat.

Potential impacts of these marine invaders include displacement of native species through competition or predation, reductions in biodiversity of coastal and estuarine habitats, and threats to fisheries and aquaculture operations. These impacts may be devastating in human terms since a large proportion of the population of Australia utilizes the coast for recreation and, indeed, a livelihood.

As yet there are no available barrier control techniques that are fully effective in preventing the entry of marine pest species into the Australian marine environment. While ports are clearly the main entry point for introduced species, these species may also colonize areas far from ports either by dispersal of eggs and larvae via natural currents, or by domestic boating activity. Monitoring the arrival and spread of introduced species is crucial to understanding how they arrive and the impact that they have, yet to date has been impossible to implement around the extensive coastline of Australia.

Members of local communities could play a vital role in this regard, since their broad geographic distribution and familiarity with natural inhabitants means that they are often the first to detect changes in local marine habitats. Recognising the need for monitoring and the value of community involvement, Environment Australia (EA) of the Federal Government, the Centre for Research on Introduced Marine Pests (CRIMP) at CSIRO, and the Australian Ballast Water Management Advisory Council (ABWMAC) are jointly funding a pilot community monitoring programme for introduced marine pests.

The community monitoring programme is to be co-ordinated by CRIMP; however environmental management agencies, industry groups, port authorities, research agencies, and established marine and monitoring networks, in addition to community groups, will have key roles. The programme aims to facilitate early detection of new invasive species, and to develop knowledge on introduced species already present in Australia, by assisting community and other groups to be active watchdogs in marine and coastal environments. Awareness raising and education will be important components of the programme, since potential participants will clearly want to understand the problem of introduced marine pests before assessing how they may participate in the monitoring programme. There are plans to develop a publicly accessible database on introduced marine pests and an information web site for the programme, in addition to other information, identification, and training materials. There will be opportunities for involvement of a broad range of groups, including divers, fishers, boaters, marine naturalists, surfers, beachcombers, and school groups, in addition to marine industry and government groups. The community monitoring programme hopes to establish a direct link for two way flow of information on introduced marine pests between CRIMP and the broader community.

Edited from an article "Community monitoring of introduced marine pests in Australia" in Aliens 6, p.14, by Karen Parsons, CSIRO, Australia.

KENYA FORESTRY RESEARCH INSTITUTE

WATCH OUT FOR THIS DREADFUL PEST OF CYPRESS

THE PEST: **Cinara cupressi** (cypress aphid)

COLOR: Brown/Yellow all over
Both winged and wingless forms occur. Young stages resemble small wingless forms.

LENGTH: 2 - 4 mm

DAMAGE: The aphid settles at the top of the crown, and colonies develop rapidly from the high fecundity of colonizing females. The crowns of infested trees turn yellow to brown under light to moderate attack. Honey dew dropping from the aphid colonies settles on lower branches and in turn grow mouldy. Infested plantations quickly develop yellow to brown tops. Trees under severe attack develop severe die-back and finally dies.

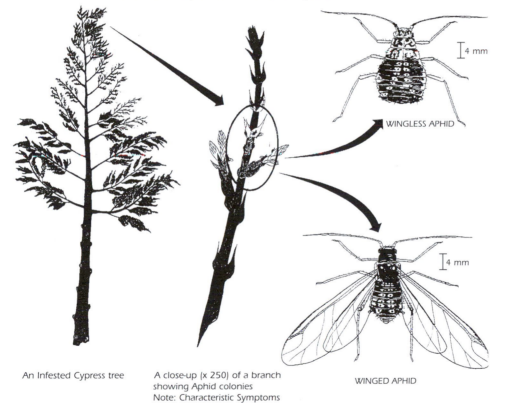

4 mm

WINGLESS APHID

4 mm

WINGED APHID

An Infested Cypress tree

A close-up (x 250) of a branch showing Aphid colonies
Note: Characteristic Symptoms of damage

WHAT TO DO
Contact: Director Kenya Forestry Reserach Institute,
P.O. Box 20412, Nairobi

Send Insect speciments sealed in bottle plus some spirit to the local Forestry Office or directly to the Director KEFRI or CAB International, P.O. Box 30148 Nairobi

CAB *International*

CASE STUDY 4.12   What Goes into a Contingency Plan?

A comprehensive contingency plan could be a large document, probably in excess of 50 pages. Very few such plans are available as yet. However, the main headings from the contents page of the draft "Contingency plan for pest animal and plant invasions on islands in the Department of Conservation, Auckland Conservancy" (March 1999) may give some flavour of what could go into such a plan.

Quick Contents

➤ Seen or suspect a rodent?

➤ Seen or suspect other new animal pests?

➤ Seen or suspect a new plant pest?

Contents

➤ Minimising the risk of rodent invasions

➤ Contingency plan for rodent invasions

➤ Minimising the risk of invasion by pest animals other than rodents

➤ Contingency plan for invasions by pest animals other than rodents

➤ Minimising the risk of plant pest invasions

➤ Contingency plan for plant pest invasions

Appendices

➤ Personnel contact list

➤ Report sheets

➤ Data sheets

➤ Location of equipment

➤ List of islands

➤ Maps of each island

➤ Lists of animals and weeds on islands

Note the "Quick Contents" right at the start of the document that should allow any person to take correct action in a panic situation. The "Contents" part of the document is also relatively brief but it must be written so that the reader does not have to find any other document to fill basic knowledge gaps. This document includes information on stopping pests getting to islands in the first place – a subject which is just as important as the contingency action but which you may wish to address in a separate document. The "Appendices" are exceedingly important and must be meticulously maintained – again, the document reader should not need to refer to any other document in relation to the important items listed in the appendices.

Of equal importance to the Contingency Plan is the involvement and commitment of all the people involved in caring for the islands. They must all understand the plan and put the protection sections into effect every day. The equipment needed for contingency action must be maintained in perfect working order and stored where the plan says it should be.

Prepared by Dick Veitch, Papakura, New Zealand.

CASE STUDY 4.13   Building a Knowledge Base for Rapid Response Action

In March 1999 a marine relative of the zebra mussel - *Mytilopsis* sp., named locally as the black striped mussel - invaded three marinas in Darwin, northern Australia. The mussel was seen to pose a threat to the environment, infrastructure and fisheries of northern Australia.

Following prompt and rigorous action the mussel was eradicated from these marinas and has not been seen since in Australia (except on the hulls of some visiting vessels) (Case Studies 3.15 and 5.23).

It was fortunate that the discovered invader *Mytilopis* sp. is closely related to the zebra mussel, so that access to relevant information on its biology, eradication and control was readily available on the internet (Sea Grants National Aquatic Nuisance Species Clearinghouse (http://cce.cornell.edu/seagrant/nansc/ SGNIS; see also Box 2.1 for more Internet-based databases) and could rapidly be used as knowledge base for a rapid response eradication programme.

A national taskforce was set up to evaluate the response to *Mytilopsis* sp. and concluded that other likely marine invaders could be equally devastating as *Mytilopsis* sp. and that it would be highly advantageous to be in the same position for a rapid response strategy against these organisms. Thus CSIRO's Centre for Research on Introduced Marine Pests has completed an extensive literature review of eradication and control approaches to marine (and some freshwater) pests. The review concentrated on taxa thought likely to pose the greatest threats to Australia. However, it was noticed that much of the eradication and control literature (especially from failed eradication attempts) has never been published. The review is available in one or more sections as downloadable .pdf files at http://www.marine.csiro.au/CRIMP/Toolbox.html.

The review will form part of an interactive "Rapid Response Toolbox" accessible through the internet and including information on the species, eradication attempts, physical and legal constraints and available experts and suppliers (the latter in Australia only). An interactive hazard analysis will be included to guide someone through a response to marine invasion identifying potential hazards and possible responses.

Edited from an e-mail from Nic Bax to Aliens discussion list, 21.9.2000. Nic Bax, Centre for Research into Introduced Marine Pests, CSIRO Marine Research, bax@marine.csiro.au

Summary

This chapter describes the management of invasive alien species including:

➤ the initial assessment of the situation,

➤ the process of identifying the species of highest priority for a management programme,

➤ detailed information on methods for eradication, containment, control, and mitigation for the various biological groups,

➤ an introduction to monitoring approaches,

➤ identification of major principles for projects,

➤ activities to secure resources,

➤ the importance of stakeholder commitment and involvement, and

➤ training in control methods.

The first step of a management programme is to assess the current situation by determining the management goal, the extent and quality of the area being managed, the invasive target species affecting the area, and the native species threatened. The management goal should be the conservation or restoration of intact ecosystems that support the delivery of ecosystem services. Eradication and control options need to be evaluated on the basis of the likelihood of success, cost effectiveness and any potential detrimental impacts.

Invasive species need to be arranged in a priority list that takes into consideration the extent of the area infested by the species, its impact, the ecological value of habitats invaded, and the difficulty of control. Species with the highest priority would be those known or suspected to be invasive but still in small numbers, species which can alter ecosystem processes, species that occur in areas of high conservation value, and those that are likely to be controlled successfully.

The four main strategies for dealing with established invasive alien species are eradication, containment, control, and mitigation. When prevention measures have failed, an eradication programme is considered to be the most effective action, because of the opportunity for complete rehabilitation of the habitat. Since eradication programmes are usually very costly and need full commitment until completion, the feasibility of eradication needs to be carefully and realistically assessed beforehand. Eradication has been achieved using mechanical, chemical and biological control, as well as habitat management. Although examples of successful eradication feature in most taxonomic groups, most success has been achieved against land vertebrates on small islands.

Containment is a specific form of control. The aim is to restrict an invasive species to a limited geographical range. The population can be suppressed using a variety of methods along the border of the defined area, individuals spreading outside this area are eradicated, and introductions outside the area prevented.

Control of invasive alien species should be planned to reduce the density and abundance of the target to below an agreed threshold, lowering the impact to an acceptable extent. The suppression of a population will reduce its competitiveness and, under optimal conditions, native species will regain ground and replace the invasive species.

The options for management of invasive alien species are very varied due to the complexity of ecosystems, species richness, and climatic regions involved. Case studies of successful programmes for the various taxonomic groups can be only guidelines, but general statements should be made with great care. Control methods are grouped by methods exploring successful attempts for the major groups.

In order to evaluate the success or failure of a programme, it is necessary to monitor changes and evaluate to what extent the targets set at the beginning of the efforts have been met. Simultaneously monitoring the impact of an eradication or control effort helps to keep the project on track and will identify negative unexpected results, giving an opportunity to change and adapt the programme to new perceptions and situations.

Some key elements for securing resources are indicated. Management of invasive alien species can be very labour-intensive, so that in some instances the costs for implementation are prohibitive. One solution to this problem can be the use of volunteer groups.

All stakeholders need to be identified in the initial assessment and integrated in the entire process of a management programme. Depending on the public perception, some species will be easier to target than others. Management of the latter group needs a broader public awareness raising element and convincing arguments. Using the media to influence the public can be a powerful tool.

A majority of successful management programmes have been implemented against agricultural and forestry pests, highlighting the importance of co-operation between different sectors. Information on success and failure must be widely disseminated via all available media. To achieve a better knowledge base, training in methods for addressing invasive alien species is needed, and a list is provided of some of the limited selection of training courses currently available.

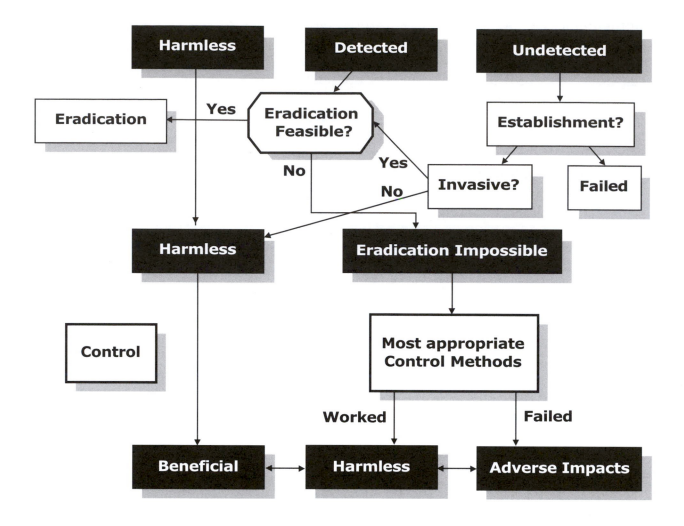

Figure 5.1 Eradication and control options after prevention has failed. After implementation of all proposed steps, the alien species will fit one of the three groups identified at the bottom of flowchart (see figure in the Toolkit Summary for the full flowchart).

5.1 Initial assessment

The first step is to determine the management goal for any invasives management project. Second, the target area needs to be defined. It may be an entire country, all or part of an island, or all or part of a reserve or conservation area.

In some instances regional projects will include more than one country and need good co-ordination between countries. Thus, it is often advisable to base an eradication or control programme of alien species on an ecosystem, which may cross political boundaries. However, sometimes the political situation might prohibit this approach.

The quality of the area, the management goal for that area, the species under threat due to invasives, and which invasive species affect the area and may adversely affect the management goal also need to be determined. The areas of highest quality for biodiversity and conservation with outstanding natural beauty, species-rich areas, and rare habitats are often protected as National Parks, with little human activity, besides tourism. The management goal for these kinds of areas will be the preservation of the natural systems, often combined with the development and maintenance of ecotourism. Smaller areas may be set aside as Nature Reserves, and here the management goal may be preservation of an ecological system, or particular parts of the ecology (e.g. flagship species); the implications for invasives management might be limited to relatively small and carefully targeted interventions.

After due consideration, it should be possible to state for a particular project what it is that is to be achieved with regard to invasives management, and how this will contribute to the overall management goals for the target area.

The management area, as defined in the management goal, has to be surveyed for alien as well as native species to assess the potential loss of natural habitat. These surveys include literature search, collection records, and actual surveys in the area. The documentation has to include the best available knowledge about the abundance and distribution of alien species, their impact on the habitat, and when justified (e.g. based on experience in a neighbouring area) a prediction of future spread and impact if not controlled. Gaps in knowledge should also be recognized. If there are earlier data available, a comparison between past and current species composition and distribution of single alien species can reveal the status and spread of species in that area. Past control actions, their success or failure, and their ecological risks should be summarized, too.

Consider the management options for each target species, using local knowledge, information from databases, published and unpublished sources. Local circumstances, such as cultural and socio-economic features of the area may affect the suitability of different options. Options for eradication, containment or control, and needs for further surveys, experimental investigations, and other research should all be evaluated. Eradication, containment and control options need to be evaluated on cost effectiveness, including possible impacts on non-target species, other possible detrimental effects, and the likelihood of success, before decisions are made.

Several of the above mentioned issues are components of risk assessment processes (see Section 3.4) that investigate potential impacts of established non-indigenous species. An assessment of spread of the introduced cane toad in Australia is provided in Case Study 5.39 "A Preliminary Risk Assessment of Cane Toads in Kakadu National Park".

5.2 Priorities for management

In this toolkit, priority setting is considered principally from the viewpoint of ecosystem and species values. However, managers should recognize that political and public support and the availability of external support may drive a pest specific project that might not be a priority from this more rigorous viewpoint.

Priorities are set in the hope of minimizing the total, long-term workload, and hence cost of an operation, in terms of money, resources and opportunities. Therefore, we should act to prevent new infestations and assign highest priority to existing infestations that are the fastest growing, most disruptive, and affect the most highly valued area(s) of the site. We also consider the difficulty of achieving satisfactory control, giving higher priority to infestations we think we are most likely to control with available technology and resources.

What follows is a stepwise approach for prioritising species and specific infestations for control. Another, more detailed, priority-setting system for weeds is presented in the *Handbook for Ranking Exotic Plants for Management and Control* (Hiebert, R.D.; Stubbendieck, J. (1993) Denver, CO: U.S. Department of Interior, National Park Service). At the national level see also the Case Studies 2.8 "Developing a Strategy for Improving Hawaii's Protection Against Harmful Alien Species", 2.11 "Summary of Australia's National Weeds Strategy", and 2.12 "The Process of Determining Weeds of National Significance in Australia".

The priority-setting process can be difficult, partly because you need to consider so many factors. It has been found that it helps to group these factors into four categories, which you can think of as filters designed to screen out the worst pests:

1. current and potential extent of the species on or near the site;

2. current and potential impacts of the species;

3. value of the habitats/areas that the species infests or may infest; and

4. difficulty of control.

The categories can be used in any order; however, we emphasize the importance of the **current extent of the species** category, and suggest it be used first. In the long run, it is usually most efficient to devote resources to preventing new problems and immediately addressing incipient infestations. Ignore categories that are unimportant on your site.

Below we suggest how species should be ranked within the four categories. If a species is described by more than one of the criteria in a given category, assign it the highest priority it qualifies for. You may assign priority in a ranking system (1, 2, 3..., n) or by class (e.g. A = worst pests, B = moderate pests, C = minor pests).

I. Current and potential extent of the species: Under this category, priorities are assigned to species in order to first, prevent the establishment of new pest species, second, eliminate small, rapidly-growing infestations, third, prevent large infestations from expanding, and fourth, reduce or eliminate large infestations. To do this, assign priorities in the following sequence:

1. Species not yet on the site but which are present nearby. Pay special attention to species known to be pests elsewhere in the region.

2. Species present on the site as new populations or outliers of larger infestations, especially if they are expanding rapidly.

3. Species present on the site in large infestations that continue to expand.

4. Species present on the site in large infestations, which are not expanding. You may have to learn to "live with" certain species or infestations that you cannot control with available technology and resources. However, keep looking for innovations that might allow you to control them in the future.

II. Current and potential impacts of the species: The order of priorities under this category is based on the management goals for your site. We suggest the following sequence:

1. Species that alter ecosystem processes such as fire frequency, sedimentation, nutrient cycling, or other ecosystem processes. These are species that "change the rules of the game", often altering conditions so radically that few native plants and animals can persist (Case Studies 5.1 "Problems Caused by Water Hyacinth as an Invasive Alien Species", 5.2 "Paper-Bark Tree Alters Habitats in Florida" and 5.3 "Chestnut Blight Changes a Forest Ecosystem").

2. Species that kill, parasitise, hybridise or outcompete natives and dominate otherwise undisturbed native communities (Case Study 5.4 "Hybridisation").

3. Species that do not outcompete dominant natives but:

➤ prevent or depress recruitment or regeneration of native species (for example, the forest understory weed garlic mustard (*Alliaria petiolata*) may depress recruitment by canopy dominants); or

➤ reduce or eliminate resources (e.g. food, cover, nesting sites) used by native animals; or

➤ promote populations of invasive non-native animals by providing them with resources otherwise unavailable in the area; or

➤ significantly increase seed distribution of non-native plants or enhance non-native plants in some other way.

4. Species that overtake and exclude natives following natural disturbances such as fires, floods, or hurricanes, thereby altering natural succession, or that hinder restoration of natural communities. Note that species of this type should be assigned higher priority in areas subject to repeated disturbances.

III. Value of the habitats/areas the species actually or potentially infests: Assign priorities in the following order:

1. Infestations that occur in the most highly valued habitats or areas - especially areas that contain rare or highly valued species or communities and areas that provide vital resources.

2. Infestations that occur in less highly valued areas. Areas already badly infested with other pests may be given low priority unless the species in question will make the situation significantly worse.

IV. Difficulty of control and establishing replacement species: Assign priorities in the following order:

1. Species likely to be controlled or eradicated with available technology and resources and which desirable native species will replace with little further input.

2. Species likely to be controlled but will not be replaced by desirable natives without an active restoration programme requiring substantial resources.

3. Species difficult to control with available technology and resources and/or whose control will likely result in substantial damage to other, desirable species and/or enhance other non-indigenous species.

4. Species unlikely to be controlled with available technology and resources. Finally, pest species whose populations are decreasing or those that colonize only disturbed areas and do not move into (relatively) undisturbed habitats or affect recovery from the disturbance can be assigned the lowest priorities.

5.3 Management strategies

We recognize four main strategies to deal with problematic non-indigenous species that have already established populations in the area under consideration: eradication, containment, control, and mitigation. Eradication is the most desirable, but often the most difficult approach. Once the establishment of an alien species is accepted as irreversible, control can be divided into containment, i.e. keeping species within regional barriers, and control in a stricter sense, i.e. suppressing population levels of alien species to below an acceptable threshold. Defining this threshold is not entirely straightforward, but it should be done before starting a control programme and it should be done in light of the management goal (Section 5.1). Ultimately the acceptable threshold relates to the level of impact on the ecosystem being invaded, but this could be expressed in terms of distribution or density or a combination of both for the invasive species. If these three management strategies cannot be employed, the last option is to try to mitigate the impact of the invasive species on native organisms and ecosystems. The strategy to find the best way in which to "live with" the introduced species is called mitigation.

Whichever management strategy is chosen, it is extremely important to choose appropriate methods to do the job and to undertake the work in the right season of the year. Most pest management methods work better at one time of the year than another and there may even be times when they are totally ineffective. Some methods will work well on one species at one time of year but for another species the same method will have to be used at a different time of year. A related topic is, which stage of the invader is most vulnerable to management methods.

5.3.1 Eradication

Eradication is the elimination of the entire population of an alien species, including any resting stages, in the managed area. When prevention has failed to stop the introduction of an alien species, an eradication programme is the preferred method of action. Eradication as a rapid response to an early detection of a non-indigenous species (Chapter 4) is often the key to a successful and cost-effective solution. However, eradication should only be attempted if it is feasible. Eradication is the type of clear-cut decisive intervention that appeals to politicians and the public, but beware of the temptations of attempting an eradication programme that is unlikely to succeed. A careful analysis of the costs (including indirect costs) and likelihood of success must be made (rapidly) and adequate resources mobilized before eradication is attempted. However, if eradication of the invasive species is achieved it is more cost-effective than any other measure of long-term control (Case Studies 4.2 "Early Detection and Eradication of White-Spotted Tussock Moth in New Zealand" and 5.5 "Eradication of a Deliberately Introduced Plant Found to be Invasive").

Eradication programmes can involve several control methods on their own or a combination of these. There are few situations where a single method is a proven eradicator of an invasive species. Therefore it is wise to plan for and use all possible methods. The methods vary depending on the invasive species. Successful eradication programmes in the past have been based on:

➤ mechanical control, e.g. hand-picking of snails and hand-pulling of weeds;

➤ chemical control, e.g. using toxic baits against vertebrates and spraying insecticides against insect pests;

➤ biopesticides, e.g. *Bacillus thuringiensis* (BT) sprayed against insect pests;

➤ sterile male releases, usually combined with chemical control;

➤ habitat management, e.g. grazing and prescribed burning;

➤ hunting of invasive vertebrates.

Some groups of organisms are more suitable for eradication efforts than others. Some methods used in past efforts are summarized below. However, it has to be borne in mind that each single situation needs to be evaluated to find the best method in that area under the given circumstances:

➤ Plants can be best eradicated by a combination of mechanical and chemical treatments, e.g. cutting of woody weeds and applying an herbicide to the cut stems (Case Study 5.6 "Eradication Programme for Chromolaena Weed in Australia").

➤ Land vertebrates. Many successful eradication programmes were carried out against land mammals on islands (Case Study 5.7 "Rabbit Eradication on Phillip Island"). The methods most frequently used were bait stations where toxic substances were offered to the invasive species, e.g. rat eradication. Bigger animals can be hunted provided the ecosystem is of an open kind with less cover to hide. A particular issue with eradication programmes against land vertebrates may be adverse public opinion, especially that of animal rights groups.

➤ Amongst land invertebrates only snails and insects have been successfully eradicated on occasion. Snails can be handpicked, whereas the commonest options to eradicate insects are based on the use of insecticides or biopesticides, usually by widespread application, or using baits or a combination of both (Case Study 5.8 "Eradication of the Giant African Snail in Florida").

➤ The use of sterile male releases, often in combination with insecticide control, has been effective on several occasions against insects, such as fruit flies and screwworm (Case Study 5.9 "Eradicating Screwworms from North America and North Africa").

➤ There are two published successful eradications of invasive species in the marine environment to date. An infestation of a sabellid worm in a bay in the USA was eliminated by hand-picking of the host (Case Study 4.4 "The First Eradication of an Established Introduced Marine Invader"), and in Australia black-striped mussel was eradicated using pesticides (Case Study 5.23 "Eradication of the Black Striped Mussel in Northern Territory, Australia"). It should be stressed that eradication in marine waters is only possible in extremely unusual circumstances that allow treatment of an effectively isolated population in a relatively contained area. Even in such events, the risks of re-invasion of the pest species is still likely to exist and will require vigilant, long-term management. In the great majority proportion of cases, eradication has been and will remain impossible.

➤ Foreign freshwater fish species have been eradicated in the past by using toxins specific to fish.

➤ Pathogens of humans and domesticated animals have been eradicated by vaccination of the respective host. In general, it seems more feasible to apply methods for eradication to the hosts (e.g. obligate alternate hosts for human diseases) rather than directly to the pathogens.

If an eradication programme is feasible, it is the preferred choice for action against an invasive non-indigenous species. The advantage of eradication as opposed to long-term control is the opportunity for complete rehabilitation to the conditions prevailing prior to the invasion of the alien species. There are no long-term control costs involved (although precautionary monitoring for early warning and/or prevention measures may be appropriate) and the ecological impacts and economic losses are diminished to zero immediately after eradication. This method is the only option that totally meets the management goal, because the invasive species is completely eliminated.

The major drawback of eradication programmes is that they may not succeed, in which case the entire investment will have been largely wasted - at most the spread of the target alien species will have been slowed. Because eradication programmes are usually very costly and need full commitment and attention until their successful completion, no eradication programme should be started unless an assessment of the available options and methods has shown that eradication is feasible. Thus, eradication should only be pursued when funding and commitment of all stakeholders are secured. Funding should be secured for a longer period than the predicted time for the eradication to allow unanticipated problems to be solved along the way and secure adequate resources for follow-up studies. Public awareness of the problems caused by the invasive species should be raised beforehand and public support sought. These steps take time, and in contrast it should also be appreciated that the more rapid the response to a new invasion, the more likely it is that eradication will succeed. These demands need to be balanced – do not let anyone tell you that decision-making in response to a new invasive alien species is easy!

A well-designed and realistic eradication approach has to be developed to achieve the required goal. In most cases, well-established populations and large areas of infestation are unsuitable for eradication programmes. Many failed attempts were highly costly and had significant side effects on non-target species, as in the case of the attempt to eradicate South-American fire ants in the southern states of the USA. The insecticide initially used proved disastrous to wildlife and cattle. The ant bait subsequently developed also had non-target effects, and proved to be more effective against native ant species than the intruder. This in fact, enhanced the populations of the non-indigenous species due to a decrease of interspecific competition with the native ant species. Finally, the eradication efforts had to be abandoned (Case Study 5.10 "Fire Ant: an Eradication Programme that Failed").

The best chances for successful eradication of most unwanted species are during the early phase of invasion, while the target populations are small and/or limited to a small area. The chances for success can be improved by identifying a period when the target species is particularly vulnerable, e.g. naturally occurring seasonal starvation periods (winter, dry season etc.) will increase the take of poison bait by mammals. Improvements in eradication technology, eradication experience elsewhere, and improved knowledge of the basic ecology of invaders will improve eradication attempts in the future. Eradication efforts have been especially successful in island situations. These can include ecological islands isolated by physical or ecological barriers, e.g. forest remnants surrounded by agricultural fields. However, the target species may survive in small populations outside an ecological island and depending upon the degree of isolation could rapidly re-invade the ecological island after an eradication campaign. The same can also be true for islands, and recolonisation by the subject of a successful eradication programme is often possible or even likely in coastal islands and archipelagos (Case Study 5.11 "Colonization Rate of Hibiscus Mealybug in the Caribbean").

For successful eradication, a rapid response against a small founder population needs to be launched as quickly as possible after detection. Part of the decision-making process is normally an assessment of whether the newly detected species is likely to be harmful in the new environment. Sometimes it is possible to anticipate the arrival of an alien species, e.g. if it is spreading within a region, in which case it may be possible to make decisions about its status before it arrives. Thus, *Chromolaena odorata* was a declared noxious weed in Queensland before it was first found there in 1994 (Case Study 5.12 "Surveying for Chromolaena Weed Infestations in Australia").

The precautionary principle could be applied, and all introductions considered as targets for eradication. If sufficient resources were available this would be the safest strategy. However, it is recognized that in most situations, prioritisation is necessary and decisions to proceed with eradication and the prioritisation of resources are dependent upon indications that the newly arrived alien is likely to be invasive, and often that it is likely to cause significant damage, especially economic damage. In future, the expectation of damage to natural ecosystems should also be a major factor in such decision-making. This assessment has to be made quickly, normally based on knowledge about the species in other countries (cf. risk assessment in Section 3.4). Databases and published literature should be checked for accounts of the species (or related species, if the species itself is not adequately known). Predicting the ability of an alien species to invade indigenous habitats and cause problems is not a precise science. Probably the best guide at present is that if a species is recognized as an invasive alien problem in one

country (particularly under similar ecological and climatic conditions), it is likely to cause similar problems in other countries (Case Study 3.23 "Invasiveness Cannot be Reliably Predicted").

Although eradication methods should be as specific as possible, the rather rigorous nature of concentrated efforts for eradication will often inflict incidental casualties to non-target species. In most cases these losses can be seen as inevitable and acceptable costs to achieve the management goal and can be balanced against the long-term economic and biodiversity benefits. When attempting eradication using toxins, it should be ensured that these are as specific as possible and that their persistence in the ecosystem is of short duration. However, some toxins unacceptable for use in a long-term control programme might justifiably be used in an eradication campaign over a short period of time.

Eradication programmes in particular need to integrate and involve all stakeholders especially the public. Management goals and the best methods to achieve them have to be discussed in an open way. Eradication of mammals, particularly those with which humans can identify, are particularly prone to opposition. Methods of killing these targets are rightly the subject of discussion and often the cause of disagreement. Animal rights groups have been known to hinder or block eradication efforts (Case Studies 5.13 "Controversy over Mammal Control Programmes " and 5.42 "Eradication of the Grey Squirrel in Italy: Failure of the Programme and Future Scenarios"). Thus, the management goal and objectives of the initiative should be written in a positive way, e.g. as an action to "rescue a poor helpless native creature from the risk of extinction brought upon it by a wild nasty invading beast" rather than just killing an invading species.

Eradication (or control) of well-established non-indigenous species, which have become a major element of the ecosystem, will influence the entire ecosystem. Predicting the consequences of the successful elimination of such species will be difficult but it must be done. The relationships (e.g. synergistic effects) of the invasive species to indigenous and non-indigenous species have to be considered. A strong carnivore-prey relationship between two invasive species points to the need to investigate the potential for combined methods to eliminate both species at the same time. Control of one species in isolation could have drastic direct or indirect effects on the population dynamics of the second species. Elimination of the normal prey may eliminate the carnivore, or it may cause it to change its behaviour and feed on native species. Elimination of an introduced carnivore is likely to allow the introduced prey to increase in numbers greatly and may cause more damage (habitat degradation, depletion of food items and competition with native species dependent on the same food) than when both were present (e.g. rabbit and red fox in Australia, both introduced from Europe). Successful eradication of a weed can also lead to negative effects in the plant community, if it is replaced by another non-indigenous plant species (Case Study 5.31 "What Can Happen When an Invasive Alien Species is Controlled"). Some of these effects on the ecosystem might not be anticipated, thus monitoring of the outcome is crucial for mitigation efforts (see Section 5.5).

By way of synthesis, basic criteria for a successful eradication programme are summarized as follows:

➤ The programme needs to be scientifically based. Unfortunately, most traits rendering species invasive make eradication efforts more difficult, e.g. high reproduction rate and dispersal ability. That means that invasive species are likely to be difficult to eliminate due to their very nature.

➤ Eradication of all individuals must be achievable. It must be borne in mind that it becomes progressively more difficult and costly to locate and remove the final individuals at the end of the programme, when the population is dwindling away.

➤ Support by the public and all stakeholders must be ensured beforehand.

➤ Sufficient funding must be secured for an intensive programme (allowing for contingencies) to make sure that eradication can be pursued until the last individual is removed. Expectations must be realistic in terms of the processes required for successful eradication programmes e.g. low visible returns for high investments late in the programme.

➤ Small, geographically limited populations of non-indigenous species are easiest to eliminate. Thus, immediate eradication is the preferred option for most species found in early detection surveys. Therefore it is crucial that the early warning programme has funds available for these actions.

➤ Immigration of the alien species must be zero, i.e. the management area must be completely isolated from other infested areas, as is the case for islands, particularly oceanic islands. Potential pathways for the species between infested areas and the management area must be controlled to prevent new invasions (cf. Chapter 3).

➤ All individuals of the population must be susceptible to the eradication technique used. If individuals learn to avoid the technique (trap-shy), they would not be susceptible to the technique and would survive. Perhaps a combination of methods more successful at high and low densities respectively would be more successful under these circumstances.

➤ Clearly defined field methods are needed which will not compromise the eradication objective – these are necessarily distinct from methods used for containment (see below). These differences must be clearly understood and quality control procedures for field practice must be put in place.

➤ Development and use of field methods will almost certainly need to be an iterative process. Implementation will needed to be monitored, followed by research to test and adapt methods to changing conditions as the point of eradication is approached. In the case of vertebrates, this research would need to be carried out on a separate population of the target species so the target population is not sensitised to new methods. This continual input of scientific knowledge and opinion has to be established from the outset, using the same personnel from the outset, but needs to be balanced with input from experienced field practitioners in a consultative and collaborative manner.

➤ Effective team management and motivation will be needed. No single person can achieve an eradication success – it has always been achieved by teamwork. A core of field and research expertise is needed to lead the eradication from the beginning to the end in order to maximize efficiency. This is particularly important to maintain the political and administrative support for the completion of the programme.

➤ A technique to monitor the species at very low densities, at the end of the programme, needs to be designed to ensure detection of the last survivors, e.g. pheromone traps installed in high densities at high risk areas. Dogs (and occasionally pigs) can be extremely successful for monitoring at low densities because of their very much greater sensitivity to the target species compared to humans or human-made detection methods. Organisms that have less obvious stages, which can survive for long periods, e.g. seed banks of weeds, need monitoring for a prolonged period (see also Case Study 5.12: "Surveying for Chromolaena Weed Infestations in Australia").

➤ A monitoring phase should be part of the eradication programme to make sure that eradication has been achieved.

➤ Methods to minimize the chances of re-invasion and early detection of the eradicated species should it re-establish need to be in place.

See also the detailed treatment by the International Plant Protection Convention (1998. Guidelines for Pest Eradication Programmes. International Standards For Phytosanitary Measures, 9. Secretariat of the International Plant Protection Convention, Food and Agriculture Organization of the United Nations, Rome, 17 pp. Also available at http://www.fao.org/ag/agp/agpp/pq/default.htm under International Standards for Phytosanitary Measures).

In the future, new technologies will be developed for use in eradication programmes. Species currently considered unsuitable for eradication may be controllable in the future. The potential use (and associated risks) of the new technologies of gene manipulation, in either the pest species or as biological control agent, are just beginning to be explored. These genetically engineered organisms may have a great potential in the future, although discussions about their safety in use are on-going. Already, the potential use of genetically engineered microorganisms for the eradication or control of introduced foxes and rabbits in Australia is being assessed.

5.3.2 Containment

Containment of non-indigenous invasive species is a special form of control. The aim is to restrict the spread of an alien species and to contain the population in a defined geographical range. The methods used for containment are the same as those described for prevention, eradication and control and are therefore not presented here in detail. Monitoring and public involvement will again be a critical feature.

Containment programmes also need to be designed with clearly defined goals: barriers beyond which the invasive species should not spread, habitats that are not to be colonized and invaded, etc. (Case Study 5.14 "Containment of the Spread of Chromolaena Weed in Australia"). In order to establish these parameters there needs to be clear understanding of why the containment is being done in the first place, e.g. to protect particular areas or habitats from invasion, to allow time to mobilize other control or eradication measures etc.

An important component of a containment programme is the ability to rapidly detect new infestations of the invasive species both spreading from the margins of its distribution, or in completely new areas, so that control measures can be implemented in as timely a manner as possible. These new infestations will initially be at very low densities, so early detection will be challenging (see Chapter 4).

The invasive species' population is suppressed using a variety of methods along the border of the defined area of containment, individuals and colonies spreading beyond this are eradicated, and introductions into areas outside the defined containment area are prevented. The distinction between containment and eradication is not always clear-cut depending upon the scale of operations considered (Case Study 5.15 "Containment vs. Eradication: *Miconia calvescens* in Hawaii").

A species most likely to be successfully contained in a defined area is a species spreading slowly over short distances. The nearest suitable habitat for the species should be preferably separated by a natural barrier, or an effective artificial barrier. The most suitable cases for containment are habitat islands without suitable connections that would allow the easy spread of invasive species. The spread of alien freshwater species between different parts of watersheds is a good example where containment may be possible.

If containment of an invasive species in a well-defined area is successful, habitats and native species are safeguarded against the impacts caused by the harmful alien species outside this area. In cases where eradication is not feasible and the range of the invasive species is restricted in a rather isolated area, containment of the species in that area can be highly effective to save the other parts of the country, even if the species is harmful in the containment area. However, a careful

analysis of the containment options, their costs and likely benefits should always be carried out.

Containing a species in a defined area will, however, need constant attention and control of the species at the border and prevention measures against spread of the species (Case Study 5.16 "Seed Movement on Vehicles: a Study from Kakadu National Park, Australia"). Thus, successful containment is difficult to achieve and involves several different costly methods.

The chances for successful containment of invasive species are relatively good for species living in freshwater habitats, e.g. fish spread limited to specific water catchment areas. Thanks to human activities, many catchment areas are connected by artificial canals that allow alien species to spread between river systems. However, canals are rather small corridors and therefore easier to control. Some species may be effectively restricted by barriers built in canals, if other pathways, such as over-land boat traffic (prevention of pathways), can be closed at the same time.

A related but different approach is exclusion, which aims to protect a sensitive area against invasive species by fencing them out. This method also often combines eradication, prevention and fencing techniques. An area of high conservation value is fenced with an animal-proof fence and if the invasive species occurs inside, it will be eradicated. This mainland-islands concept is very effective in supporting crucial populations of endangered species, if eradication of the invasive species within the containment is possible but eradication on a large-scale is not feasible.

5.3.3   Control

Control of non-indigenous invasive species aims for the long-term reduction in density and abundance to below a pre-set acceptable threshold. The harm caused by the species under this threshold is considered acceptable with regard to damage to biodiversity and economy. It is not always clear what this level should be set at in order to achieve the management objective. Research to establish what indigenous biodiversity is at risk and how much of the invasive species' impact can be tolerated may need to be carried out.

Suppression of the invasive population below that threshold can tip the balance in favour of native competing species. The weakened state of the invasive species allows native species to regain ground and even further diminish the abundance of the alien species. In rare cases this might even lead to extinction of the non-indigenous species (especially combined with habitat restoration efforts to support native species and put intact natural systems back in place), but this is clearly not the principle goal of control efforts.

If prevention methods have failed and eradication is not feasible managers will have to live with the introduced species and can only try to mitigate the negative impacts on biodiversity and ecosystems. All control methods, with the exception of classical biological control, which is self-sustaining, need long-term funding and commitment. If the funding ceases, the population and the corresponding negative impacts will normally increase, perhaps leading to irreversible damage.

Since, in the short-term, control seems to be a cheaper option than eradication, it is often the preferred method. Funding and commitment do not need to be at such high levels as for eradication programmes, and funding can be varied between the years depending on the perceived importance of the problem, political pressure, and public awareness. However, the lower recurring costs are deceptive, because in the long run effective control is more expensive in total than a successful eradication campaign.

Mechanical, chemical and biological control, habitat management, and a combination of methods are all used successfully in controlling population levels of invasive species. In many cases a cost-effective combination of appropriate measures may be put together in a sustainable way so as to minimize side effects. This is integrated pest management as developed in the agricultural and forestry sectors, based on long and bitter experience of chemical insecticide dependence (for method descriptions see the following chapters). In many countries it is now the preferred national pest management strategy for sustainable crop production, and many of the principles can be applied for management of alien species by the environment sector.

Successful control may be easiest to achieve in areas of lower density of the invasive species. Such control will immediately mitigate the impact of the invasive species, allowing a relatively intact ecosystem to recover from the impacts of the alien species. Successful control in these areas will rapidly show a positive effect on biodiversity, and where it is on the edge of the range of the invasive species, the spread of the alien will be limited.

Population levels of invasive species of any organism group can be controlled to a certain extent using suitable available control methods. The degree of success will vary with different organisms, the ecosystem, the duration of effort, the restoration effort, etc. It is one of the lessons of agricultural pest management that the optimum pest management strategy is often location specific and must be tested and fine-tuned for different areas. This should be kept in mind with regard to all the methods and Case Studies described below.

5.3.4 Mitigation

If eradication, containment, and control are not options or have failed in managing an invasive alien species, the last resort is to "live with" this species in the best achievable way and mitigate impacts on biodiversity and endangered species.

Mitigation as used in this context differs from containment and control in that the activity undertaken does not directly affect the invasive species in question but rather focuses on affected native species. Mitigation is most commonly used in the conservation of endangered species and can be approached at various levels. At its simplest and perhaps most extreme form it could mean the translocation of a viable population of the endangered species to an ecosystem where the invasive species of concern does not occur or, in the case of a rehabilitated system, no longer occurs.

In the case of vertebrates, however, more often it involves some minor alterations in the behavioural patterns of the desired species. This most commonly entails the conditioning of animals to use specific, often artificial, nesting and/or roosting sites that are by their nature or design inaccessible to the invasive alien species, or artificial feeding sites/dispensers in instances of feeding competition or habitat degradation.

It should be noted that mitigation can be labour intensive and costly and is often seen as an intermediate measure to be taken in tandem with eradication, containment or control for immediate mitigation efforts to rescue a critically endangered native species from extinction. However predator-proof nest boxes have been successful in many instances pertaining to bird conservation, e.g. the Mauritian Kestrel and the Seychelles Black Parrot (Case Study 5.41 "Invasive Species Mitigation to Save the Seychelles Black Parrot").

5.4   Methods

There are a large number of specific methods to control invasive species. Recognizing the highly complex nature of invasives ecology and the importance of local conditions, general statements about suitable control methods for groups of alien species, in specific habitats or world regions should be approached with great caution. Precise predictions of the behaviour, spread, and impacts of non-indigenous species introduced into new environments are not available, because too many of the parameters used to describe the situation are no more than informed guesses. In many cases even the taxonomic status of the invasive species is uncertain. However, descriptions are available of methods used to control certain species and their effectiveness under specific environmental factors. These experience-based reports are essential for invasives management and need to be made increasingly available, for example in databases accessible through the Internet. The goal of anyone involved in invasives management should be to use the best practices available and to disseminate information to serve the higher goal of preserving the earth's biodiversity and mitigating problems caused by invasive organisms on a worldwide scale.

A wealth of information is available from experience in pest control in agriculture and forestry. Prevention and control of pests in these sectors has been going on for a long time and is of great value as an information base for management of invasive species in natural habitats. Many key agriculture and forestry pests are non-indigenous species – as are the plants used in agriculture and forestry – and have been managed for many decades using specific methods for prevention, eradication and control. Moreover, most of the facilities and services in place to deal with foreign species at the borders or within a country have been initiated against agriculture and forestry pests, and are managed by the ministry responsibly for that sector. The quarantine facilities and other related services should be used and expanded to address environmental pests as well. Apart from the many similarities between those alien species invading natural ecosystems and those affecting agriculture and forestry, it should be remembered that many species affect both areas. For example, the problems caused by water hyacinth are multi-faceted, affecting the mandates of several ministries (Case Study 5.1 "Problems Caused by Water Hyacinth as an Invasive Alien Species") and thereby creating additional problems for the organization of its prevention, containment or control.

The first step to the development of a successful control strategy for an invasive species is therefore to check literature and databases to accumulate as much information as possible about management options for this species. Successful methods used under similar conditions, i.e. in similar habitats and climates, should be tested. Use of less than optimal methods is not recommended. The most successful invasive species control has been achieved with species-specific methods, which also have the least impact on non-target species. In some instances, such as highly degraded habitats without any native species left, a more general method is acceptable. In these cases a broad-spectrum herbicide, or bulldozing the ground, has limited negative effects on native biodiversity. However, in less disturbed areas, in particular nature reserves etc., the use of a species-specific method is highly recommended.

Non-target casualties can generally be expected while carrying out such control measures. This can be a significant public relations issue where the casualties are non-target vertebrates. A small level of non-target casualty may have to be accepted to achieve the objectives, but this must not be allowed to reach unacceptable levels. When control or eradication is successful, the reduced impact of the alien species on the native biodiversity normally outweighs the casualties. In most cases populations of the native species that suffered losses during the control efforts, will rebound following the removal of the invasive species (Case Study 5.17 "Reptile Recovery on Round Island").

In each country, different tools are already available to control invasive species and there are differing rules on the use of these tools, e.g. pesticide registration, quarantine issues and the legal framework. It is therefore necessary for the conservation sector in each country to build an information file, which recognizes the legislative, technical and best practice for control of different species. The

GISP database (and existing web sites and published documents – see web sites and linkages therein of Box 2.1) is seen as an important information-sharing source for development of these documents. This information file can be developed as work on each species is considered.

Important points to consider are:

➤ All legal requirements related to management of invasive species. Some of these may be hidden in health and safety legislation.

➤ Support from groups who will appreciate the benefits of the project, such as the scientific community, animal welfare groups and others.

➤ Best methods that have been used for this target species in other countries.

➤ The types of herbicides, baits and equipment that are readily available in the country and the ways by which further supplies can be obtained.

In most cases the best practice to manage an invasive species may involve a system of integrated management tailored for the species and the location. Thus, it is important to accumulate the available information, assess all potential methods, and use the best method or combination of methods to achieve the target level of control. Always bear in mind that managing an invasive species is not the management goal, but only one tool in the process to achieve a higher goal, such as habitat restoration, preservation of an undisturbed ecosystem, re-installation of the natural succession rate and time, etc. These intact areas can provide sustainable use of ecosystem services to humans. As previously indicated these higher goals need to be clearly defined and quantified, and when planning a control programme it is worth setting a time scale for achieving these goals, with benchmarks and perhaps indicators if possible.

The successful control of the population of an invasive species itself can have indirect effects on native species, the ecosystem, and the entire local biodiversity. The potential effects of reducing or eradicating the invasive species in a habitat should be evaluated beforehand and measures taken to ensure that these effects are solely positive. For example, removal of an aggressive invasive plant from a site might need to be accompanied by planting of indigenous species to fill the gaps, to prevent these gaps being filled by other unwanted plants (cf. Case Study 5.31 "What Can Happen When an Invasive Alien Species is Controlled").

Plant control may involve: manual methods (e.g. hand-pulling, cutting, mowing, bulldozing, girdling); herbicides; release of biological control agents; controlled use of grazing or browsing animals; prescribed fires; flooding; planting competitive native species and other land management practices.

Land invertebrate control may involve traps (e.g. light traps, pitfall traps, pheromone traps), mechanical/physical means (e.g. handpicking, removal and destruction of host species), insecticides, biological control (e.g. fungi, other insects), and other specialized means (e.g. mass release of sterile males – see Case Study 5.9 "Eradicating Screwworms from North America and North Africa").

Land vertebrate control may involve trapping, shooting, baiting, biological control, contraceptives or sterilization (Case Study 5.44 "Eradication Programmes against the American Mink in Europe"). Lizard and snake control is a little known subject area that needs more research, although considerable work is being developed in relation to control of the brown tree snake in the Pacific.

Control of **pathogens** often focuses on hosts rather than measures directly orientated against the pathogen species. In some cases the hosts are eliminated – this is a preferred choice when the hosts are non-indigenous as well – in others, including diseases of humans and domesticated animals, the hosts are vaccinated. Resistance of the host can also be induced or intensified. If vectors are a part of the pathogen's lifecycle, vector management should be considered.

There is not much experience with control of **marine** bioinvasions, but for example an invasive seastar has been removed mechanically by hand-picking (Case Study 5.19 "Mechanical and Chemical Control of Seastars in Australia are Not Promising"), and another was eradicated by applying pesticides (Case Study 5.23 "Eradication of the Black Striped Mussel in Northern Territory, Australia"). Whereas biological control using parasitoids, predators or pathogens is used successfully in the control of many types of alien species, it has as yet never been attempted in a marine environment. However, research to investigate biological control options against some marine invaders is currently under way. Management of invasive species in marine environment is apparently more difficult than in terrestrial areas for various reasons.

There are considerable gaps in knowledge regarding taxonomy of marine species. Related to this is the lack of information available on the natural ranges for most marine organisms. Thus, discovering the origin and solving the question as to whether a species exhibiting invasive behaviour is non-indigenous or actually native proves to be very difficult in many cases. If there is no doubt about the invasive organism, control is probably still difficult to achieve in the more open marine environment where it is more difficult to apply targeted control measures. Many marine species are adapted to quickly changing conditions and have evolved mechanisms to spread throughout the ecologically suitable range. Many sessile and semi-sessile organisms have pelagic larvae, highly capable of long-distance dispersal with the help of water currents. Thus, naturally occurring borders are fewer than in terrestrial systems, making control and eradication efforts more difficult. For this reason, prevention is generally considered the principle defence against marine invasive species, which are distributed by one principal pathway, the ship.

Organisms in **freshwater** habitats can be controlled with mechanical, chemical, and biological measures and habitat management. Aquatic weeds can be harvested when floating on the surface, pulled out when rooted, or sprayed with herbicides. Biological control has been particularly effective against several water weeds in different parts of the world (Case Study 5.26 "Biological Control of Water Weeds"). Fish-specific poison has been used in the eradication of several fish invasions. Another control option for fish is recreational or industrial fishing. Mosquito larvae and pathogens vectored, and by extrapolation other freshwater insects, can be controlled by spraying chemicals or biological pesticides onto the infested water. The community of a freshwater system can be influenced by changes in the water quality and quantity in favour of native species.

Trained staff are an important component of all management methods. For some methods, in some countries, the level of training is determined by law (e.g. herbicide training and a certificate for use may be required).

5.4.1 Mechanical control

Mechanical control can be carried out by directly removing individuals of the target species either by hand or using tools. In many cases introduced pests can be controlled or even eradicated in small-scale infestations by mechanical control, for example hand-pulling weeds or handpicking animals. An advanced method of mechanical control is the removal of plants by specifically designed tools and even machines, such as harvesting vehicles for water hyacinth infested lakes and rivers. In some cases of very persistent plants and depending on the area, e.g. on large open areas like pastures, bulldozing may be necessary (recommended in the described circumstances for autumn olive – *Elaeagnus umbellata* – Randall, J.M.; Marinelli. J. (1996) *Invasive Plants, Weeds of the Global Garden*. Brooklyn Botanic Garden, Handbook #149, Brooklyn, New York. 99 p. http://www.gardenweb.com/bbg/plant.html).

Mechanical control can be used in both eradication programmes and as a means for controlling densities and abundance of invasive species. Basically, all organisms can be removed mechanically one way or another. However, available information needs to be screened and control should be carried out or supervised by trained staff to choose and apply the most effective way. Eradication will often only be achieved in small areas.

Mechanical control is highly specific to the target and non-target effects are mostly restricted to disturbance by human presence. The downside of the method is the fact that it is always highly labour-intensive. In countries where human labour is costly, the use of physical methods is limited mainly to volunteer groups. Most manual work is expensive and has to be repeated for several years to remove all individuals. For weeds whose seeds can be dormant in the soil for a long period, monitoring for that potential dormancy period after eradication is necessary. The method can be effective when the population of the invader is still small and the

population is limited to a small area. Weeds that grow vigorously from cut plant parts or multiply vegetatively are more difficult to control.

Invasive **plants** can be cut, hand-pulled or removed by specific tools (for some simple tools see http://tncweeds.ucdavis.edu/tools.html). Larger plants can be uprooted, with the aid of tools, such as winches, if necessary. The effectiveness of this technique will vary considerably depending on the response of the weed (Case Study 5.18 "Conservation Management Areas in Mauritius"). Plant parts of some species, left in contact with soil may survive and grow, for example Japanese Knotweed (*Fallopia japonica*), an invasive alien in Europe and North America, will regenerate from rhizome fragments of less than 1 gram. If there is no information available about the plant's response to uprooting, some simple tests should be carried out to discover its effectiveness, and ways to treat the residues, e.g. composting or burning the uprooted material.

Repeated cutting of a woody weed may eventually drain the resources stored in the root system and kill the plant. In many cases combined cutting of the plant and painting the stem with a systemic herbicide proves to be more efficient. Specialized cutting tools that will apply a pesticide as they cut have been tested. Mowing of herbs and grasses may lead to the same result, when the plants are not adapted to heavy grazing. Annuals are especially susceptible if mown shortly before setting flowers, because they will have used up most of their root reserves to produce the buds.

Girdling can kill trees; cutting with a knife through the cambium of a tree trunk and removing 5 cm of bark will interrupt the flow of nutrients and kill the plant. Girdling alone may not suffice for rapidly killing those species where the water and nutrient movement are not restricted to the outermost layer of the trunk, but an application of herbicide will speed up the process.

Large, visibly obvious **invertebrate** species, such as snails, can be handpicked. For control of most insect species one is dependent on traps, which are more or less specific to insect groups or species-specific using pheromones. Sedentary species such as scale insects or mealybugs can be killed by destroying their food plants, for example a containment programme against the newly arrived hibiscus mealybug in Trinidad involved cutting and burning infested plants, with follow up applications of pesticides.

Trapping and shooting can be considered the "mechanical" or "manual" way of dealing with invasive **vertebrates**. Recreational hunting of game can be effective in keeping populations down to an acceptable level and can be a money source for other management activities in the area. This is a rare case where control does not involve costs, but earns money. It does, however, give rise to the concern that the invasive species then becomes a valuable commodity that should be preserved in order to continue to generate this income. Furthermore, there are many instances where recreational hunting will not reduce the target population sufficiently. Similarly, recreational hunting can be counter productive due to amateur hunters creating a shy target population and not being skilled enough to reduce target species down to desired densities. Also, depending on the species, recreational hunters may select only mature trophy males as targets; this will have little or no impact on the reproductive capacity of the species. In order to reach the pre-determined target population level, it may be necessary to employ professional hunters. Using animals such as dogs, which can be specifically trained to target individual invasive species, can be extremely successful in combination with shooting and other forms of control.

Fencing is another option for containment of species, either fencing the species in a certain area or fencing off ecologically valuable land. One obviously needs to be sure that the invasive species is not present on both sides of the fence.

An example of an eradication programme by handpicking of a **marine** bioinvader is summarized in the Case Study 4.4 "The First Eradication of an Established Introduced Marine Invader", but this approach is generally of limited applicability in marine ecosystems. Mechanical control has been used against seastars but was not very effective (Case Study 5.19 "Mechanical and Chemical Control of Seastars in Australia are Not Promising").

Perhaps the only mechanical control method against **pathogens** is to eradicate or control the vector or the host, e.g. felling of diseased trees.

Aquatic weeds can be harvested (Case Study 5.20 "Mechanical Control Methods for Water Hyacinth") - as are their terrestrial relatives. Certain fish species are of high commercial value and/or popular with sport fishing. There are financial parallels identifiable between fishing and hunting in respect to popular species. Economically viable harvesting of invasive species, however, generates the risk of providing the incentive for some individuals to spread the invasive species to new areas, not yet colonized.

5.4.2 Chemical control

Chemical pesticides, including herbicides and insecticides, have been developed to meet the markets for control of pests in agricultural production, and elimination of disease vectors. Development, testing and registration of a new compound is a very expensive process, and few products are likely to be developed specifically to address environmental targets. Nevertheless, products developed for the agriculture and human health sectors are available to those trying to control invasive species, and can be used to decrease population levels of invasive organisms below a threshold of ecologically tolerable impact.

In the past, extensively used broad-spectrum herbicides such as DDT had massive detrimental impacts to the environment as well as human health, but today these are banned in most countries, and there are more specific products on the market with fewer negative non-target effects. Some insecticides, such as those based on chemical structures similar to insect hormones, can also be specific to target groups of insects.

Major drawbacks are the high costs, the necessity of repeating an application, and the impacts on non-target species. An additional problem very clearly demonstrated in agriculture and human disease vector control, is that repeated use of pesticides provides the selective pressure which enables many target species to evolve increasingly effective resistance to these chemicals. In response either the dose has to be increased or a different group of pesticides has to be used, usually further increasing the control costs.

There is also the possibility that indigenous peoples will oppose the use of toxins on their land, for example where toxins may accumulate in sub-lethal levels in non-target species that may be an important food source for indigenous peoples. This latter concern is mainly true of persistent pesticides such as modern anti-coagulants and the now largely obsolete organo-chlorine compounds; the available pesticide registration data should clarify how serious this risk is. One example of opposition to the use of a persistent anti-coagulant has occurred in the Far North of New Zealand. Here the local tribe of Maori (the indigenous people of Aotearoa - New Zealand) have opposed the use of brodifacoum used to kill rats that prey on native giant land snails. Their reasoning has been that the poison may persist in the environment and "taint" the "purity" of the land and reside in species used for food such as feral pig (*Sus scrofa*), ironically itself introduced into New Zealand by Europeans settlers and a threat to the giant land snails.

Selection of a pesticide to control an invasive species begins with a determination of effectiveness against the target and all appropriate non-target species that might come in contact with the chemical, either directly or through secondary sources. Additionally, the environmental half-life, method of delivery, means of reducing non-target species contact, demonstration of efficacy, and collection of data to ensure compliance with environmentally safe use (as set out by the regulatory bodies in the country where it will be used) must be evaluated. Most countries require pesticides to be registered for specific uses. Once identified, tested, and registered, a pesticide can allow the rapid control of a target species over large areas, and as a result reduce the need for personnel and costs for the more traditional methods such as traps and barriers.

Widely used application methods for **herbicides** include treatments of the bark of young trees or applying herbicide into the wounds created by girdling or cutting. This cut-stump application method, mentioned already in the section on mechanical control, is very effective against many woody plants. Herbicide can also be applied directly to the leaves of the invasive species by using a sponge or wick, but a less specific method is foliar spraying of infested areas (Case Study 5.21 "Chemical Control of *Miconia calvescens* in Hawaii").

Similarly, **insecticides** can be sprayed selectively on infested plants or plant parts or indiscriminately over a large area. Application should always be as focused as possible on the pest, e.g. spraying of the attacked plant part, at the most susceptible time for the target, and limiting the use to the efficient dose, in order to minimize side-effects on other species.

Pesticides are used against **vertebrates** mainly in baits, e.g. bait stations for rats. Before using bait, small-scale experiments and observations can be carried out to determine which non-target species might take the bait. With some ingenuity, it may be possible to develop bait stations to give easy access to the target species but prevent, as far as possible, other species from entering it. Obviously, a more target specific bait station is easier to design for an ecosystem with no species similar to the target species (Case Study 5.22 "Overview of Successful Rat Eradications on Islands").

Chemical substances are used to mitigate **diseases** in humans and animals. Disinfection of water and surfaces capable and suspected of disease transmission are treated with disinfectants to kill pathogens before entering their hosts.

Chemical treatment offers one of the few options for control of **marine** invasive species, although its potential is limited (Case Study 5.19 "Mechanical and Chemical Control of Seastars in Australia are Not Promising"). In Northern Territory, Australia an eradication programme using pesticides was successfully carried out against a marine invasive organism, the black striped mussel, *Mytilopsis* sp. (Case Study 5.23 "Eradication of the Black Striped Mussel in Northern Territory, Australia").

Herbicides (e.g. glyphosate and 2,4-D) have been used extensively around the world as a quick and effective means of controlling weeds in **freshwater** environments. However, since they are non-selective and more difficult to apply directly to the target plant in water, they are more likely to cause harm to non-target species. The fish poison rotenone (for overlap with biopesticides see Section 5.4.3) is frequently used to control fish species in ponds and other small water bodies. This method is efficient for the eradication of species, but the non-selective character limits its use for large-scale infestations.

There is a large literature on formulation, application and use of pesticides particularly for the control of insects and weeds (Box 5.1 "Some Reference Sources on Chemical Pesticides").

5.4.3   Biological control

Biological control is the intentional use of populations of upper trophic level organisms commonly referred to as natural enemies, or naturally synthesised substances against pest species to suppress pest populations. Biological control can be split up in several approaches grouped under two headings: those that are self-sustaining and those that are not. Methods that are not self-sustaining include:

➤ Mass release of sterile males to swamp the population with males which copulate with the females without producing any offspring in the next generation – see Case Study 5.9 "Eradicating Screwworms from North America and North Africa".

➤ Inducing host resistance against the pest. This approach is particularly relevant to agriculture where plant breeders select (or create) varieties resistant to diseases and insects.

➤ Biological chemicals, i.e. chemicals synthesised by living organisms. This category overlaps with chemical control and whether to list a particular method in one or the other category is a question of definition, e.g. while applying living *Bacillus thuringiensis* (BT) is without doubt a biological control option, to which group the use of the toxins stored in BT belong could be debatable (Case Study 5.25 "*Bacillus thuringiensis*, the Most Widely Used Biopesticide"). Other examples of chemicals in this group are rotenone, neem and pyrethrum, extracted from plants.

➤ Inundative biological control using pathogens, parasitoids or predators that will not reproduce and survive effectively in the ecosystem. Large-scale or mass releases of natural enemies are made to react quickly to control a pest population.

Self-sustaining biological control includes:

➤ Classical biological control. At its simplest, this is the introduction of natural enemies from the original range of the target species into new areas where the pest is invasive. Invasive alien species are often controlled in their indigenous range by their natural enemies, but are usually introduced into new environments without these natural enemies. Freed of their parasitoids, parasites and predators alien species often grow and/or reproduce more vigorously in the country of introduction. Natural enemies for introduction are selected on the basis of their host specificity to minimize or eliminate any risk of effects on non-target species. The aim is not the eradication of the invasive alien, but to reduce its competitiveness with native species, hence reducing its density, and its impact on the environment.

➤ Augmentation of enemies under pest outbreak conditions for an immediate control, when the enemy can reproduce in the new environment. The control agent is reared or cultured in large numbers and released.

➤ Habitat management (see Section 5.4.4) can enhance populations of native predators and parasitoids, e.g. release/replant of native alternate hosts and food resources.

The most important of these for management of invasive alien species is classical biological control. Conservation managers are coming to realize that this method, if used following modern protocols such as the International Plant Protection Convention's *Code of Conduct for the Import and Release of Exotic Biological Control Agents* (see Box 5.2), provides the safest and most cost efficient approach to solve many invasive alien species problems.

In comparison with other methods, classical biological control is, when successful, highly cost-effective, permanent and self-sustaining. It is ecologically safe due to the high specificity of the agents used. The main disadvantages are the lack of certainty about the level of control that will be achieved, and the delays until the established agents achieve their full impact. However, with a potentially very positive benefit:cost ratio, the benefits of classical biological control normally outweigh the drawbacks and it represents the cheapest and safest option to date.

There has been quite some debate in recent years about the safety of classical biological control, particularly with regard to the potential of introduced biological control agents to have adverse effects on non-target organisms. In particular, some of the introductions made over 50 years ago were of generalist predators, including vertebrates such as mongooses and cane toads, and these did have severe adverse effects on non-target populations, including species of conservation importance. Such species would not be used today in biological control, and some of them are good examples of invasive alien species causing serious problems. However, today the safety standards of biological control are very rigorous. It is a normal requirement (e.g. IPPC Code of Conduct) to assess the specificity of all agents proposed for introduction. This involves extensive laboratory and field screening tests. An informed decision can then be made by the appropriate national authority taking into consideration the potential for any effects on non-target organisms.

While biological control is highly recommended to control an established population of an invasive alien species, the theory of natural population regulation underlying the principle of biological control does not anticipate eradication with this method. In a successful biological control programme, the invasive species' population will be reduced to an acceptable level, but the populations of prey/host and predator/parasitoid will remain present in a dynamic balance. Biological control is particularly appropriate for use in nature reserves and other conservation areas because of its environmental-friendly nature and the prohibition of pesticide use in many such areas (Case Study 5.24 "Biological Control of an Insect to Save an Endemic Tree on St. Helena").

Box 5.2 "Some Reference Sources on Biological Control" provides an entry point to the literature on biological control.

Pheromone traps, based on chemicals produced by the target species to attract other members of the same species, are species- or genus-specific in most cases and allow the selective collection of the target species. Occasionally species may be controlled effectively by using high densities of traps, particularly in a small or restricted area. Thus, if the pheromone is readily and cheaply available in large amounts, the release of high doses of the pheromone can interfere with mate location and mating. If the air is filled with the pheromone the insects are not able to detect and find a partner. This method is only feasible for small infestations and is mainly used in orchards, greenhouses and similar conditions.

Generally pheromone traps are more effective when used to monitor the presence or abundance of a species. For instance, traps can be used for early detection of high-risk species. This may enable a rapid response action to attempt eradication or containment. Traps can also be used to monitor the density of pest species, so that when the catches reach a certain threshold other control measures are triggered. The progress of an eradication programme can also be followed by monitoring the density (and later the lack) of the target species.

Biopesticides

Biopesticides are biological pesticides based on beneficial insect and weed pathogens and entomopathogenic (i.e. insect-killing) nematodes. Pathogens used as biopesticides include fungi, bacteria, viruses and protozoa. Produced, formulated and applied in appropriate ways, such biopesticides can provide ecological and effective solutions to pest problems.

As yet all product development has been directed towards control of pests having direct economic impact, particularly for the control of pests of agriculture, forestry and horticulture (caterpillars, locusts, various beetles, weeds), medical and nuisance pests (mosquitoes, blackflies and flies).

Most types of biopesticides are relatively specific to their target pests, and many are very specific. It is this specificity which makes their use attractive compared to broad-spectrum chemical pesticides. The most widely available and used biopesticides are various formulations of *Bacillus thuringiensis* (known as 'Bt'), which can be used to control the larval stages of Lepidoptera (caterpillars), and selected Coleoptera (beetles) and Diptera (e.g. mosquitoes and flies) (Case Study 5.25 "*Bacillus thuringiensis*, the Most Widely Used Biopesticide").

Entomopathogenic nematodes are increasingly available in specialized niche markets, such as horticulture and are used to kill selected invertebrate pest targets.

Fungi for control of specific weeds ("mycoherbicides" or "bioherbicides") have been available for some time, and the development of new ones is increasingly routine (see e.g. International Bioherbicide Group http://ibg.ba.cnr.it/). These products are usually host specific either due to the physiology of the fungus, or because of the way they are used. This makes their use attractive in many situations, but also means that the market is small, making them commercially less attractive than traditional herbicides. Nevertheless, a niche market exists, and could be developed to address specific conservation needs to control invasive alien plants, as part of a management programme. For example, the development and use of mycoherbicide products to be used for stump painting in the control of plants such as *Rhododendron ponticum* in Europe is under consideration.

Fungi for control of insects is also a relatively new research area, but products are now coming onto the market, notably Green Muscle, a formulation of *Metarhizium anisopliae* for control of locusts and acridid grasshoppers (http://www.cabi.org/bioscience/biopesticides.htm). See Case Study 2.15 "Mauritius and La Réunion Co-operate to Prevent a Sugar Cane Pest Spreading" for an example of practical use of a fungus against an alien beetle pest of sugar cane.

Thus, at the moment biopesticides will be of value for management of invasive alien species where there is a suitable product already available, but for the future the technology and expertise is available to develop target-specific products for control of particular invasive species.

Pathogens for control of vertebrates

Not only can pathogens be used as biopesticides but there are also opportunities to use them against vertebrates, e.g. against the brown tree snake playing havoc with Guam's ecology or the release of myxoma virus (myxomatosis) and calicivirus (rabbit haemorrhagic disease) against rabbits in Australia. Snakes differ markedly from birds and mammals in susceptibility to various diseases. Viral or bacterial pathogens capable of killing or weakening only the brown tree snake (and thereby reducing its population) are an attractive objective. Unlike more traditional interventionist techniques, a disease might spread with little human assistance and remain effective for years. Potential pathogens must be carefully screened for risks to other animals and humans. Thus, pathogens, like chemical insecticides require significant preliminary testing and verification prior to use, although these costs might easily be offset by rapid and widespread distribution in the brown tree snake population once released. Controlled and extensive laboratory experiments involving virologists, ecologists, and pathologists are required to test pathogens. Work is underway at the Guam National Zoological Park to determine the susceptibility of brown tree snake to a viral pathogen from zoo disease outbreaks and other sources.

Biological control of freshwater and marine targets

The opportunity to use biological control against plants, invertebrates and vertebrates are described above. Classical biological control against water weeds has been particularly promising and has produced several success stories (Case Study 5.26 "Biological Control of Water Weeds").

No biological control project has being attempted against a marine invader to date, though studies on the suitability of several parasites against different organisms are underway, e.g. specific parasitic castrators of crabs (Case Study 5.27 "Possible Biological Control for European Green Crab").

Biological control of plant diseases

Biological control of plant diseases is still a young science. Many plant pathogens colonize parts of the plant that are initially free of microorganisms. Successful biological control in such circumstances depends on rapidly colonising these plant areas with non-pathogenic antagonists competing for the space. The principal antagonists used are saprotrophic fungi and antibiotic-producing bacteria. The biological control agent will ideally outcompete the pathogen. This concept is altogether a rather different approach than the biological control projects against weeds, invertebrates and vertebrates. In some cases less virulent strains of the same pathogen species can be used to replace the virulent strain physically or by transmission of the traits of the less virulent strain to the virulent one.

5.4.4 Habitat management

Prescribed burning

In certain environments the practice of prescribed burning can change the vegetation cover in favour of native plant species, thereby decreasing population levels of weeds. Prescribed burning is particularly appropriate for restoring or maintaining fire-adapted or fire-dependent species and natural communities. Many invasive plants are not adapted to fire; thus, ecological burning may be an effective tool for controlling these species. However, land managers must first determine if fire is a natural component in the plant community in question and if prescribed fire can be expected to help meet site goals.

Fire has been used quite frequently to manage invasive alien species in the USA, for example to eradicate Australian pine (*Casuarina equisetifolia*) in pine forests and other fire-tolerant communities in the USA (Case Study 5.28 "Control Methods for Australian Pine Include Prescribed Burning"), but less frequently elsewhere. Spot treatment is also possible, for example, early in the growing season baby's-breath (*Gypsophila paniculata*) can be burned with a hand-held propane torch. On the other hand, it should be remembered that growth of some invasive alien plants, such as garlic mustard (*Alliaria petiolata*) in woodland of North-East USA are stimulated by fire.

Only trained and experienced people should undertake prescribed burning due to the many health and safety risks involved. Smaller infestations can be controlled with the aid of a flame-thrower. The risks of a large-scale fire limit the use of these tools, especially in dry climates. Given these ecological and logistical challenges, prescribed burning may not be an appropriate method if considered for invasive species control only. Clearly, it is best suited to a site where restoration and maintenance of fire-dependent or fire-tolerant communities are primary conservation goals.

Grazing

Habitat management with grazing mammals can be a suitable option to obtain the desired plant cover. This method works best where the plants that are to be preserved are adapted to grazing, i.e. they are either adapted to high populations of large herbivorous mammals or prevalent in human-made habitats such as pastures and heathland. On the other hand unmanaged grazing often favours alien plants, as grazing can preferentially remove native vegetation leaving alien plants, especially toxic species, to grow under reduced competition. This twofold enhancement leads sooner or later to a monotypic stand of an alien plant, e.g. leafy spurge infestations in the USA.

Changing abiotic factors

Most invasions of non-indigenous species are caused or at least favoured by human disturbance of the ecosystems. In these cases a mitigation of negative impacts by the invasive species could be achieved by changes in the human behaviour that has led to the invasion. An example would be a change in the quantity of nutrients and/or water available for plants, which would alter the plant community. In some cases invasive aquatic organisms can be controlled by improving the water quality, addressing eutrophication and pollution problems, or even changing the quantity of water, e.g. draining or a water level regime adverse for the invasive species.

Hunting and other use of non-indigenous species

Continuous hunting can be used to control exotic species, such as deer, originally introduced for hunting purposes. There are two approaches: commercial hunting principally for meat and recreational hunting. Both approaches can generate income for the landowner and/or the state. Some exotic species are both comparatively easy to hunt and are favoured species for hunters, and so should be straightforward to manage by hunting, but conversely more wary species or those less preferred by hunters are less likely to be effectively managed.

Problems encountered trying to control an alien species through hunting usually relate to land ownership and the distribution of the invasive species. Some species spread into suburban areas where hunting is not allowed. Significant groups of the human population, particularly in developed countries, find hunting morally unacceptable, and so it may decrease in popularity, thus allowing alien species formerly controlled by hunting to explode in numbers.

Many other invasive species can be eaten or have edible fruits, which can be exploited for human consumption or as fodder for domesticated animals. In many parts of the world with high human density invasive plants are esteemed also for their production of highly valued firewood or other uses. A high percentage of introduced fish and crustacean species make a good meal, thus recreational as well as industrial fishing is certainly helping to control invasive fish populations.

However, in the promotion of an alien species as a food resource lurks the danger of providing an incentive for individuals to spread the alien species to as yet uninfested areas, or breed them in captivity, from where they may eventually escape. This issue has to be evaluated on case-by-case basis in order to estimate the potential danger and benefits.

5.4.5   Integrated Pest Management (IPM)

Over the last 30-40 years, pest management in agricultural and forestry systems has evolved from using a single control method targeted at a specific pest or group of pests on a crop, to an increasingly holistic approach. The development of the concept of integrated pest management (IPM) was probably the first and largest step, when it was recognized that different pest control methods could be used in combination to achieve the desired pest control, and that this desired level of control was within acceptable thresholds, not simply the lowest pest population that could be achieved. Subsequently workers in the field went on to recognize that pest management should deal with all the pests affecting a crop in an integrated manner – solving the problem of one pest may simply generate new and often worse problems with other pests. More recently, extension staff will talk increasingly in terms of integrated crop production, identifying that ultimately it is the yield and profit of the harvest that is the objective, and that pest management is a means towards maximising these outputs. Finally it is appropriate to recognize that pest management methods (notably the use of chemical insecticides) can have adverse effects on humans and their environment, and the costs of these need to be considered; hence farm output and profits need to be optimised in light of these costs, rather than simply maximized.

The parallel with regard to the management of invasive alien species is clear. Initially one may think in terms of controlling individual invasive alien species using one control method. Then it will become apparent that greater success can be achieved by using control methods in an integrated way (see for example Case Study 2.15 "Mauritius and La Réunion Co-operate to Prevent a Sugar Cane Pest Spreading"). Next the effects of successful management of one species need to be interpreted and may well lead to the need to manage a suite of invasive alien species and habitat management practices to achieve the desired goal. Finally, as has already been stressed, the ultimate goal of preservation (or increase) of indigenous biodiversity in the conservation area being managed, needs to be kept to the forefront of planning and monitoring invasive alien species management. An advanced form of IPM mentioned in the preceding chapter is habitat management involving several methods to meet the goal of preserving or restoring the ecosystem and its functions.

Combining methods, such as those already described in this chapter, will often provide the most effective and acceptable control. The integration of methods based on ecological research, regularly monitored, and co-ordination will almost always achieve the best results in managing an invasive species' population and reaching the overall goal. This integrated process needs an assessment of the situation and probably an experimental part for the best practice to establish protocols for the management of the invasive species (Case Study 5.29 "An IPM Research Programme on Horse Chestnut Leafminer in Europe").

The process of control can be complicated, involving several different tactics in combination or in sequence (Case Study 5.30 "Integrated Management of Water Hyacinth"), or it may be that at its simplest just one method applied using a simple decision rule can suffice to achieve the objective required. A simple method combining two approaches that is frequently used against woody weeds uses mechanical and chemical control. The mature plants are cut down and a systemic herbicide is immediately applied onto the top of the living inner bark layer of the stump. Glyphosate and Triclopyr are the most widely used herbicides for this application form.

Implementing IPM programmes is dependent on a large number of variables, so that no general recommendation can be given for any taxonomic group. Indeed it is the conventional wisdom of IPM in agriculture that although the broad outline of an IPM system can be prescribed, the local variation in these factors means that the detailed programme will end up being location-specific, evolving over time. Thus any IPM programme for an invasive alien species has to evolve based on knowledge available on the invasive organism, the ecosystem invaded, the climatic conditions, and other native and alien species thriving in the same habitat (Boxes 5.3 "Some Reference Sources on IPM " and 5.4 "Some Internet Reference Sources on IPM").

We should point out one rather significant difference between applying IPM (or its more holistic derivatives) against indigenous pests in an agricultural situation, and applying it against invasive alien species, particularly in a conservation situation. IPM is often, and ideally, based on gaining the maximum benefit from those aspects of crop production that do not need a specific intervention. For example, one standard approach, which is used effectively in many agricultural systems, is to minimize or eliminate the application of insecticides so that the natural enemies normally found in the agro-ecosystem are preserved, and can have the maximum impact on the key pests. However, when dealing with invasive alien species whether in conservation areas or agriculture, one reason why they are invasive is because they have become established without their specialized natural enemies from their area of origin, and the generalist natural enemies that are present, are not effective. This is partly why IPM strategies are normally location specific.

Because of this lack of effective indigenous natural enemies, it may be necessary to implement a successful biological control programme before an effective IPM strategy based on the introduced natural enemies can be put in place.

5.5   Monitoring and follow-up

As has been pointed out earlier, in order to evaluate the success or failure of the management efforts, it will be necessary to monitor aspects such as the population of the invasive species, the condition of the area under consideration, and changes in species composition and importance.   A management programme is not complete unless it is based on thorough preparation, persistent efforts during the programme, and follow-up studies.   Control activities, whether they involve eradication efforts, control actions taken, or taking no action at all, must be monitored over the period of the programme.   The targets set at the beginning will help to evaluate the success or failure of the programme.

The overarching goal is preservation or restoration of natural habitats to a predetermined level.   To evaluate progress, a subset of targets should be set up which are on the way to the final goal.   These targets for success may be the removal of the invasive species if the option chosen was eradication, but if it was control then the criteria for success may be a measure of some other feature, such as the return of a plant or an increase in abundance of a bird.   These criteria for success will help decide whether the programme is succeeding in controlling the pest and preserving or restoring the species and communities wanted.

For more information on how to design an effective monitoring programme, check out for example the web site of the Ramsar Convention on Wetlands (at www.ramsar.org).

It is worth pointing out that monitoring the numbers of a pest species killed or removed is a measure of the work being done but is not a measure of success of the project.   Success of the project can only be measured by monitoring numbers of the pest species that remain, and ultimately the condition of the ecosystem they are in. It should not be assumed that removing an invasive alien species from an ecosystem will automatically lead to the return of the indigenous flora and fauna.   Often this will happen (see for example, Case Study 5.17 "Reptile Recovery on Round Island"), but in other cases, removal of one alien species may simply open the way for colonization by another (Case Study 5.31 "What Can Happen When an Invasive Alien Species is Controlled").   Monitoring of the impact of control actions needs to be put in place, preferably starting with small-scale activities to verify the impact of control operations, and if the results are not as expected, the management plan may need to be reconsidered and adapted in light of this new knowledge.

In most cases successful eradication programmes need to be accompanied by prevention measures against re-colonization by the removed species and early warning systems should be put in place to detect colonisers early. A new infestation of the successfully eradicated species can be wiped out swiftly when detected early by using the appropriate eradication method, because the knowledge of the negative impact of the invasive species and the experience in controlling the species is established and will be supported by the stakeholders.

5.6 Project management

This section is inevitably generic, and restricted to general statements of principles by which aspiring project managers should be guided. Making detailed and concrete recommendations about the "human" side of project management would not be very useful, as the toolkit must operate at a global generic level, and human aspects will often be very project-specific and dependent on local circumstances. These aspects may need to be addressed in regional or national adaptations of this toolkit.

A well-managed alien control programme should have a clear project plan, including:

➤ Thorough preparation using databases and other information available.

➤ Stakeholder involvement.

➤ A timeline and milestones.

➤ An adequate budget of money and time (be very clear about the number of worker-days and operating costs required to complete each milestone): Use a matrix chart or, where capacity allows, use project management software (bottom up budgeting). An eradication programme suspended due to money constraints will be a failure, wasting time and funding!

➤ Be aware and 'up front' about the need for prolonged commitment to manage many invasives in the long term.

➤ Regular monitoring to see whether milestones are reached (make sure this is properly budgeted) (cf. Section 5.5).

➤ Make sure monitoring data is analysed rapidly to allow adaptive management/re-planning if control measures are not working (see for example Case Study 5.31 "What Can Happen When an Invasive Alien Species is Controlled").

➤ Make data available to other countries and regions with similar problems.

A distinction can be made between site-led and species-led programmes. Identifying which programme is the suitable one for the situation confronted can help in focusing on the real objectives. Site-led programmes aim to protect a specific site or area from all or most damaging invasives – see for example, Case Studies 5.18 "Conservation Management Areas in Mauritius", and 5.33 "Social and Environmental Benefits of the Fynbos Working for Water Programme". Species-led programmes aim to minimize damage by one or a few key invasive species, usually across an entire island or region. Biological control is always species-led – see e.g. Case Study 5.24 "Biological Control of an Insect to Save an Endemic Tree on St. Helena".

In some cases the land on which the invasive species spread may need to be leased or other agreements have to be taken to ensure management of invasives on private land.

5.7 Securing resources

This is a generic problem for many types of activity and certainly not restricted to the prevention and management of alien species. Securing resources will also normally be a very location-specific activity, and it will be difficult to generalize in a way that would be useful to a majority of countries. Accordingly, we try to provide some pointers in this section, and highlight a few aspects that may be especially relevant for conservation work, particularly invasive alien species problems.

In most cases a programme will be generated as the result of a manager presenting a proposal for funding, either internally or externally to a government department, a donor agency, a foundation, a non-government organization or some other source of resources. A good proposal will:

➤ Spell out the benefits in clear terms.
➤ Maximize the capacity-building opportunities.
➤ Seek and involve appropriate international partners to raise funds.
➤ Be clear on the timeframe and the budget needed.
➤ Be honest about the uncertainties.
➤ Be reviewed to make sure that it is clear before submission.

One option to increase the chance of acceptance of the proposal is to seek appropriate international partners to raise funds, e.g. CABI, IUCN, WWF and developed country national programmes. International partnerships are often needed to address the challenges of invasive alien species (Case Study 5.32 "Development of a European Research Programme on Horse Chestnut Leafminer").

Possible sources of funding particularly appropriate for management programmes to preserve biodiversity in developing countries include: GEF (Global Environment Facility), developed country aid budgets, WWF, etc. It would be beneficial if this type of information could be compiled and made more widely available.

If the case for management of alien species can be linked directly to economic or social issues, this is likely to make it more attractive for government support and funding. The Working for Water Programme in South Africa (Case Study 5.33 "Social and Environmental Benefits of the Fynbos Working for Water Programme") and linking ecotourism with cat and rat eradication in the Seychelles (Case Study 5.34 "Ecotourism as a Source of Funding to Control Invasive Species") are good examples of what can be achieved.

5.7.1 Use of volunteers

Pest control, particularly for weeds, is often very labour-intensive and thus, depending on the local costs, very expensive. In some cases, obvious results – a feeling of having achieved a small contribution towards a better world - can be rewarding by themselves to many people, so that they volunteer their participation in the management of ecosystems. In many cases local residents may be interested, but the possibility of recruitment from overseas should not be ignored, especially where this can be self-sponsored. Small tropical islands may be attractive to volunteers from developed countries with environmentally conscious populations and traditionally miserable climates. The Mauritian Wildlife Foundation has been particularly successful in this approach (Case Studies 5.35 "Use of Volunteers" and 5.38 "The Use of Local Part-time Volunteers to Help Restore a Nature Reserve on Rodrigues"), and a similar scheme has been adopted in Singapore (Case Study 5.43 "Students Help to Restore a Rainforest by Weeding"). On the other hand community groups dealing with invasive species need to be trained. Training, staff and travel costs can exceed the benefits from an ill-guided volunteer group. Hence, the groups have to be led through a successful initiative.

However, it should also be remembered that using volunteers and maintaining acceptable quality control is often difficult and requires experienced supervisors. Human resource skills in professional wildlife and habitat managers in the field do not necessarily go hand in hand!

5.7.2 Tapping of other resources

The profile of an invasive alien species project can be raised in public opinion, for example by selecting a popular species that will benefit from the project and linking this species with the project. Fluffy, cute and cuddly species suffering as the result of a bioinvasion are of course particularly effective (just as controlling similar species when they are invasive alien species is especially difficult – Section 5.8). Newspapers, radio stations, and television shows need to be positively influenced to gain a more widespread increase in status for, and interest in, the project. Commercial companies may welcome opportunities to sponsor certain prestigious projects. Chemical companies may provide free pesticides for special initiatives, whereas other companies may provide free tools and equipment. Airlines may provide free flights or discounts for travel.

In Mauritius, sugar cane companies provide free labour in the off-season for weeding fenced plots in natural forest to remove alien species (Case Study 5.18 "Conservation Management Areas in Mauritius").

Donations may be provided from the public and other fund-raising organizations. Charitable organizations may give support. Imagination in the search for resources can be very helpful and sponsorship may be found in unexpected ways. On the other hand spending too much time on searching for resources may distract from the main work and be at the expense of managing invasives!

In some countries job creation schemes may be created to generate affordable labour to help the public good. Procuring unemployed people for invasives management (and other tasks) is currently under discussion in developed countries, where labour is otherwise expensive, and full employment potentially possible and politically desirable. Unemployed people could be given a task related to invasives management in exchange for the support that they receive from the government. These initiatives and the tasks allocated would provide opportunities for social contact and education about invasives as well.

5.8 Engaging stakeholders

A stakeholder is any person or organization who will be affected, or think they will be affected, positively or negatively by the species or sites planned to be managed by the proposal. This may include funding agencies, landowners, tenants, conservation bodies, potential employees, national and local government, relevant NGOs, pressure groups, members of the public, and so on.

While the project goals and milestones are developed, the stakeholders need to be identified and integrated into the process from the start (Case Study 5.40 "Community-based Aboriginal Weed Management in the 'Top End' of Northern Australia"). Stakeholders should be consulted about the goals of the project and the activities that need to be undertaken to reach the goal. The process should be open and all questions and concerns raised by the stakeholders should be addressed. Where opinions differ, and agreement with or between some stakeholders cannot be achieved, some modification of the programme should be considered if this will lead to agreement leading to healthy co-operation.

If stakeholders are not involved with the process from the onset, later on groups can argue that they did not know about the project and may actually stop the programme in progress. The people or organizations involved do not necessarily have to be of high importance to make a point. Several eradication efforts came to a halt after interventions from little-known organizations, especially when mammals (cute, fluffy, cuddly and charismatic syndrome) or attractive trees were being controlled (Case Study 5.13 "Controversy Over Mammal Control Programmes").

It is helpful if the pest species can be portrayed as bad, causing serious damage to the natural environment. Often this is easy, but plants with pretty flowers, like old man's beard (*Clematis vitalba*) in New Zealand, or attractive leaves, like *Miconia calvescens* on Pacific islands, or furry mammals, like the brush-tailed possum in New Zealand, can be loved by humans. A determined and sustained effort through the media, possibly over many years, emphasizing the other side of the story, forest trees killed by old man's beard, *M. calvescens* clothing entire hillsides, and possums killing native trees, is needed.

Education of the public and support from the media is crucial to achieve a successful invasive species programme (cf. Section 2.4 for detailed information) – see Case Studies 4.6 "Public Awareness and Early Detection of *Miconia calvescens* in French Polynesia", 5.14 "Containment of the Spread of Chromolaena Weed in Australia" and 5.36 "Using the Media to Create Awareness and Support for Management of Invasive Species: the Seychelles Experience".

If an awareness raising campaign using the media is successful, the programme will receive public attention and respect. If the public can actually be involved, people may start to identify with the project, try to help to solve the problem, and will be proud to be part of a successful campaign (Case Study 5.37 "Community Participation in Control of Salvinia in Papua New Guinea"). This is the basis for the use of volunteers, which can be crucial to a project where significant human-power is needed. New Zealand has developed a system for distribution/redistribution of biological control agents involving the land-owner and other interested people, informed through clear illustrated brochures showing how to identify and translocate the species. The public can also be educated about the negative impacts of invasive species on the native biodiversity and ecosystem functioning. People get involved by pulling up weed outliers, and even killing introduced possums and cane toads on roads by running them over. Noxious weeds acts and similar regulations should be the foundation for any independent behaviour of this type, because doing harm to other wildlife is prohibited. Likewise hunting can be encouraged or even a bounty paid on captured or killed invasive species. An eradication campaign on Zanzibar (East Africa) was carried out paying a bounty for every house crow killed. This system worked quite well as a control option, involved local people and provided some with cash income. However, some people started to take advantage out of the bounty approach by rearing the crows. This is, of course, counter productive, and the bounty had to be stopped.

Involvement and support by the media for control of invasive alien species is crucial for the programme to be successful. It has been suggested that access to media is often much easier in small island developing states where the radio and television stations need reports for their programmes (see e.g. Case Study 4.6 "Public Awareness and Early Detection of *Miconia calvescens* in French Polynesia"). Stories about invasive species told with some humour and tension make excellent news. Project plans can be promoted via the media, telling the people what the plans are and why it is necessary to do something about invasive alien species. Co-ordination of press releases and public events can be very effective in raising the profile in the public eye.

5.9   Training in invasives control methods

There have been many successful control programmes for terrestrial and freshwater weeds, terrestrial arthropods and vertebrate pests in island and mainland settings around the world.   Many of these, particularly those against weeds and arthropods, have been in an agricultural or forestry setting.   Successful control programmes against other groups of invasive species are much less common, but examples have been highlighted in this text.   Written accounts of many of these successful programmes are available (in journals, proceedings, books, and government reports) and can be used as examples in training programmes.   Good written accounts of unsuccessful programmes are also useful learning tools, though regrettably, but understandably, they are often not written.

It will be useful to consult with the people who conducted successful programmes elsewhere.   Consultation with outside experts is especially useful and recommended in difficult or little known situations, such as dealing with marine invasions – in some instances the invasive species will be little known in the scientific community and there may be no control method known.   In any case, if targets or situations not previously controlled are encountered, techniques will have to be developed and tested as the programme proceeds.

The actual tools for invasives management are those used for agricultural pest management, and training in these areas is widely available.   Organized training in how to specifically apply these tools to invasives management, i.e. development of strategy and plans, is limited.

The variation in national capacity and types of invasive species makes it difficult to be prescriptive about what training is appropriate for what groups of staff.   Some possibilities are set out below:

➤   Incorporating "on-the-job" activity works well and requires little extra resources.

➤   In-country courses; these can be specifically targeted to ecosystems or species, using overseas experts.

➤   Overseas international courses.   These are likely to be generic rather than specific.   They are likely to be orientated towards management of existing pest species (particularly economic pests) rather than management of invasive species (particularly newly invasive species).   We are not aware of many courses targeted on management of marine invasive species or vertebrates, but other potentially useful courses are listed in Box 5.5 " Some Short Training Courses Relevant to Invasive Species Management".

➤ Study tours and attachment to successful programmes can be very valuable.

➤ Higher education; various BSc and MSc level courses are relevant to aliens management, although again most will be targeted towards management of economic pests rather than of environmental pests.

5.10   Training for planners and managers

Many land managers are not confident in their own planning ability and would benefit from going through an exercise of writing a pest control plan with others. If someone in-country has experience they can share their experience and train or assist other land managers to write plans. If not, it may be helpful to have an experienced person from another nation to give an introduction course for managers.

BOX 5.1   Some Reference Sources on Chemical Pesticides

Books and papers

Brent, K.J.; Atkin, R.K. (eds.) (1987) *Rational Pesticide Use*. Cambridge University Press, 348 pp.

Bull, D. (1982) *A growing problem: pesticides and the third world poor*. Oxfam, Oxford, 192 pp.

Carroll, N.B. (1999) Pesticide development in transition: crop protection and more. In: *Emerging technologies for Integrated Pest Management*. Eds.: G.C. Kenedy & T.B. Sutton., American Phytopathological Society Press, St. Paul, Minnesota, USA, pp 339-354.

Green, M.B.; Hartley, G.S.; West, T.F. (1977) *Chemicals for crop protection and pest control*. Pergamon Press Ltd., Oxford, 296 pp.

Hofstein, R.; Chapple, A. (1998) Commercial development of biofungicides. Pp. 77-102 in Hall, F.R.; Menn, J.J. (eds.) *Biopesticides: use and delivery*. Humana Press, Totowa, NJ, USA.

Loevinsohn, M.E. (1987) Insecticide use and increased mortality in rural Central Luzon, Philippines. *The Lancet* (June 1987), 1359-1362.

Matteson, P.C. (2000) Insect pest management in tropical Asian irrigated rice. *Annual Review of Entomology*, **45**, 549-574.

Matthews G. A. (1992) *Pesticide application methods*. 2nd Edition. Longman Scientific and Technical, Harlow, Essex, UK, 405 pages.

Mumford, J.D.; Knight, J.D. (1997) Injury, damage and threshold concepts. Pp. 203-220 in Dent, D.R.; Walton, M.P. (eds.) *Methods in Ecological and Agricultural Entomology*: CAB International, Wallingford.

Tomlin, C.D.S. (ed.) (1997) *The Pesticide Manual*. 11th Edition, British Crop Protection Council, Bracknell, UK, 1606 pp.

van Emden, H.F.; Peakall, D.B. (1996) *Beyond silent spring: integrated pest management and chemical safety*. Chapman and Hall, London, 322 pp.

Way, M.J.; van Emden, H.F. (2000) Integrated pest management in practice - pathways towards successful application. *Crop Protection*, **19**, 81-103.

Web-sites:

http://piked2.agn.uiuc.edu/wssa/ Herbicide information from the Weed Science Society of America.

http://www.cdpr.ca.gov/ California Department of Pesticide Regulation; useful links page.

http://ace.ace.orst.edu/info/extoxnet/ The EXTOXNET InfoBase provides a variety of information about pesticides.

http://www.ianr.unl.edu/ianr/pat/ephome.htm University of Nebraska's Pesticide education resources.

http://refuges.fws.gov/NWRSFiles/InternetResources/Pesticide.html The US Federal Interagency Committee for the Management of Noxious and Exotic Weeds links to sources of information about pesticides on the internet.

BOX 5.2 Some Reference Sources on Biological Control

Books

Bellows, Jr., T.S., T. W. Fisher, L. E. Caltagirone, D. L. Dahlsten, C. Huffaker and G. Gordh (1999) Handbook of Biological Control. Academic Press.

Cameron, P.J., R.L. Hill, J. Bain and W.P. Thomas (eds.) (1989) *A Review of biological control of invertebrate pests and weeds in New Zealand 1874 to 1987*. CABI, Wallingford, UK, 424 pp.

Clausen, C.P. (ed.) (1972) *Introduced parasites and predators of arthropod pests and weeds: a World review*. USDA-ARS Agriculture Handbook 480, 545 pp.

Cock, M.J.W. (ed.) (1985) *A review of biological control of pests in the Commonwealth Caribbean and Bermuda up to 1982*. Commonwealth Institute of Biological Control, Farnham Royal, UK, 218 pp.

DeBach, P. (1964) *Biological control of insect pests and weeds*. Chapman & Hall, London, 844 pp.

Harley, K.L.S. and I.W. Forno (1992) Biological control of weeds a handbook for practitioners and students. Inkata Press, Melbourne, Australia, 74 pp.

Hokkanen, H.M.T. and J.M. Lynch (1995) *Biological control benefits and risks*. Cambridge University Press, UK, 304 pp.

Julien, M.H. and M.W. Griffiths (eds.) (1998) *Biological control of weeds. A World catalogue of agents and their target weeds*. Fourth edition. CABI, Wallingford, UK, 223 pp.

Kelleher, J.S. and M.A. Hulme (eds.) (1981) *Biological control programmes against insects and weeds in Canada 1969-1980*. Commonwealth Agricultural Bureaux, Franham Royal, UK, 410 pp. (Update in preparation).

Van Driesche, R.G. and T.S. Bellows, Jr. (eds.) (1993) *Steps in classical arthropod biological control*. Entomological Society of America, Lanham, Maryland, 88 pp.

Waterhouse, D.F. and K.R. Norris (1987) *Biological control Pacific prospects*. Inkata Press, Melbourne, Australia, 454 pp. (and two supplementary volumes).

Waterhouse, D.F. (1994) *Biological control of weeds: southeast Asian prospects*. Australian Centre for International Agricultural Research, Canberra, 302 pp.

Waterhouse, D.F. (1998) *Biological control of insect pests: southeast Asian prospects*. Australian Centre for International Agricultural Research, Canberra, 548 pp.

International Guidelines

International Plant Protection Convention (1996) Code of Conduct for the Import and Release of Biological Control Agents. International Plant Protection Convention, Food and Agriculture Organization of the United Nations, Rome. 19 pp. Also available at http://www.fao.org/ag/agp/agpp/pq/default.htm under International Standards for Phytosanitary Measures and at http://pest.cabweb.org/PDF/BNI/RA36.PDF

Information Journal

Biocontrol News & Information: http://pest.cabweb.org/Journals/BNI/Bnimain.htm

Selected websites

http://gnv.ifas.ufl.edu/~iobcweed/ The Nearctic Regional Section of the International Organization for Biological Control's Biological Control of Weeds Working Group.

http://ipmwww.ncsu.edu/biocontrol/biocontrol.html Biological Control Virtual Information Center.

BOX 5.3   Some Reference Sources on IPM

Books

General

Conway, G.R. (ed) (1984) *Pest and pathogen control: strategic tactical and policy models*. John Wiley and Sons Inc., 487 pp.

Dent, D. (2000) *Insect pest management*. 2nd edition. CABI, Wallingford, UK, 410 pp.

Matthews, G.A. (1984) *Pest management*. Longman, UK, 231 pp.

Mengech, A.; K.N. Saxena and H.N.B. Gopalan (eds.) (1995) *Integrated pest management in the tropics: Current status and future prospects*. John Wiley & Sons, Chichester, 171 pp.

Metcalf, R. and W. Lucmann (1994) *Introduction to insect pest management (3rd Edition)*. Wiley and Sons Inc., 650 pp.

Morse, S. and W. Buhler (1997) *Integrated pest management: ideals and realities in developing countries*. Lynne Reinner, London, UK, 170 pp.

Norton, G.A. and J.D. Mumford (1993) *Decision tools for pest management*. CAB International, Wallingford, UK. 264 pp.

Sindel, B.M. (ed.) Australian Weed Management Systems. RG and FJ Richardson, Meredith, Victoria, Australia, 506 pp.

Farmer Participatory IPM

Chambers, R.; A. Pacey and L.A. Thrupp (1989) *Farmer first: farmer innovation and agricultural research*. Intermediate Technology Publications, London, 240 pp. (*This book is followed by a series. Titles include: Farmers' Research in Practice; Lessons from the field, Joining Farmers' Experiments, Let Farmers Judge.*)

Scoones, I.; J. Thompson; I. Guiit and J. Pretty (1995) *Participatory learning and action*. Intermediate Technology Publications, London.

Scarborough, V.; S. Killough; D. Johnson and J. Farrington (1997) *Farmer led extension: concepts and practices*. Intermediate Technology Publications, London, 214 pp.

Data Bases on CD Rom

Crop Protection Compendium, available from CAB International, Wallingford, UK (http://www.cabi.org/Publishing/Products/CDROM/Compendia/CPC/index.asp).

BOX 5.4   Some Internet Reference Sources on IPM

Organizations

http://www.cabi.org/ CAB International, and http://www.cabi.org/BIOSCIENCE/ CABI Bioscience
http://www.cgiar.org/ Consultative Group on International Agricultural Research (CGIAR)
http://www.fao.org/ FAO; http://www.fao.org/waicent/faoinfo/agricult/agp FAO Plant
     Protection Division; http://www.communityipm.org/ FAO Programme for community IPM in Asia
http://nbo.icipe.org International Centre for Insect Physiology and Ecology (ICIPE)
http://www.nri.org Natural Resources Institute (NRI); http://www.nri.org/Themes/ipm NRI
     Integrated Pest Management Programme
http://www.pan-international.org/ Pesticides Action Network (PAN) International
http://www.ucdavis.edu/ University of California, Davis; http://www.ipm.ucdavis.edu/
     Statewide Integrated Pest Management Programme
http://www.ifgb.uni-hannover.de/extern/ppigb/ppigb.htm Institute for Plant Diseases,
     University of Bonn, Germany.
http://www.worldbank.org/ World Bank;
http://wbln0018.worldbank.org/essd/essd.nsf/rural+development/portal/ World Bank Rural
     development and agriculture; http://www-esd.worldbank.org/extension/ WB Extension

IPM Networks

http://www.nri.org/ipmeurope IPM Europe
http://www.nri.org/ipmforum/ IPM Forum
http://www.ipmnet.org/brochure.html/ IPM Net
http://www.ipmnet.org/about.html/ The Consortium for International Crop Protection
http://www.nysaes.cornell.edu/ipmnet/ne.ipm.primer.html/ Integrated Pest Management in the
     Northeast USA

Databases and Resource Centres

http://bluegoose.arw.r9.fws.gov/NWRSFiles/InternetResources/IPMPage.html/ Blue goose
http://www.ent.iastate.edu/List/ Entomology Index of Internet Resources
http://refuges.fws.gov/NWRSFiles/InternetResources/IPMPage.html Federal Interagency
     Committee for the Management of Noxious and Exotic Weeds
http://www.nysaes.cornell.edu/ent/hortcrops/ Global Crop Pest Identification and Information
     Services in Integrated Pest Management (IPM)
http://www.cf.ac.uk/insect/index.html Insect Investigations Ltd. School of Biosciences, Cardiff
     University, UK
http://www.ipmnet.org:8140/DIR/ IPM Net Database of IPM Resources
http://www.igc.org/panna/resources/resources.html/ Pesticide Action Network Resources
http://pest.cabweb.org/ Pest CABWEB
http://ipmworld.umn.edu/ Radcliffe's IPM Textbook
http://ipmwww.ncsu.edu/cipm.html/ Virtual centre for IPM

Management

Integrated Approach to Invasive Species Management (4 days). US Fish & Wildlife Service, National Conservation Training Center, Rt 1, Box 166, Shepherds Grade Road, Shepherdstown, WV 25443, USA. Contact Chris Horsch, Aquatic Resources Training; tel. ++ 1 304 876 7445; fax ++ 1 304 876 7225; e-mail chris_hrosc@fws.gov; web-site http://training.fws.gov/catalog/ecosystem.html#invasives

Plants

Biological control of tropical weeds (two weeks every other year). The Centre for Pest Information Technology & Transfer. Contact the Short Course Co-ordinator, CPITT, The University of Queensland, Brisbane QLD 4072, AUSTRALIA; Fax ++ 61 7 3365 1855; E-mail Courses@CPITT.uq.edu.au

Control Methods for Invasive Plants (one-day course). New England Wild Flower Society 180 Hemenway Road, Framingham, MA, USA; Contact NEWFS Education Dept., e-mail: registrar@newfs.org. More Information: http://www.newfs.org/courses.html#special

Introductory (one day) course for reductions of assisted weed spread and associated management systems. Environmentally Aware Contractors, Geelong, Victoria, Australia. Contact Bruce Dupe, EAC Project Officer; tel. ++ 3 5267 2104; e-mail bjd@primus.com.au

Many of the courses related to weeds in Australia are listed in the "Weed Navigator". It is available for sale through the Weeds CRC at crcweeds@waite.adelaide.edu.au

Management of invasive alien plants in the UK (one day courses targeted for individual plant species). Contact Dr Lois Child, Centre for Environmental Studies, Loughborough University, Loughborough, LE11 3TU, UK; e-mail: L.E.Child@lboro.ac.uk

Aquatic Weed Control Short Course (one week, annual course). University of Florida/IFAS, Office of Conferences and Institutes (OCI), PO Box 110750, Building 639, Mowry Road, Gainesville, FL 32611-0750, USA. Tel. ++ 1 352 392 5930; fax ++ 1 352 392 9734; e-mail: bamt@gnv.ifas.ufl.edu; web-site http://www.ifas.ufl.edu/~conferweb/#upcoming.

Noxious Weed Management Short Course (one week, annually in April). Contact: Celestine Duncan (Co-ordinator), Weed Management Services, P.O. Box 9055, Helena, MT 59604, USA; tel. ++ 1 406 443 1469; e-mail weeds1@ixi.net.

Invertebrates and plants

Biological pest management (4-5 week annual course). CABI Bioscience, Silwood Park, Ascot, Berks, UK. Contact the training officer, Mark Cook; tel. ++ 44 1784 470111; fax ++ 00 1491 829100; e-mail m.cook@cabi.org; web-site: http://www.cabi.org/BIOSCIENCE/training.htm

Marine

Short training course on invasive marine species of San Francisco Bay and the central California coast (a one-off course but could be held again in response to demand). University of California, Davis and other institutions. Contact Edwin Grosholz; Department of Environmental Science and Policy; One Shields Avenue; University of California, Davis; Davis, CA 95616, USA; tel. ++ 1 530 752-9151; fax ++ 1 530 752-3350; e-mail tedgrosholz@ucdavis.edu; web-site http://www.des.ucdavis.edu/faculty/grosholz.html.

CASE STUDY 5.1   Problems Caused by Water Hyacinth as an Invasive Alien Species

Water hyacinth (*Eichhornia crassipes*), native to South America, but now an environmental and social menace throughout the Old World tropics, affects the environment and humans in diverse ways. Most of these are detrimental, although some are beneficial or potentially useful. Many of these effects are due to its potential to grow rapidly and produce enormous amounts of biomass, thereby covering extensive areas of naturally open water.

A most striking and little understood effect of water hyacinth is on aquatic plant community structure and succession. Water hyacinth replaces existing aquatic plants, and develops floating mats of interlocked water hyacinth plants, which are colonized by several semi-aquatic plant species. As succession continues, floating mats dominated by large grasses may drift away or be grounded. This process can lead to rapid and profound changes in wetland ecology, e.g. shallow areas of water will be converted to swamps. In slow-moving water bodies, water hyacinth mats physically slow the flow of water, causing suspended particles to be precipitated, leading to silting. The reduced water flow can also cause flooding and adversely affect irrigation schemes. Water hyacinth acts as a weed in paddy rice by interfering with germination and establishment. Water hyacinth is reported to cause substantially increased loss of water by evapo-transpiration compared to open water, although this has recently been challenged. Displacement of water by water hyacinth can mean that the effective capacity of water reservoirs is reduced by up to 400 m^3 of water per hectare, causing water levels in reservoirs to fall more rapidly in dry periods. Water displacement, siltation of reservoirs and physical fouling of water intakes can have a major impact on hydroelectric schemes. Water hyacinth mats are difficult or impossible to penetrate with boats, and even small mats regularly foul boat propellers. This can have a severe effect on transport, especially where water transport is the norm. Infestations make access to fishing grounds increasingly time consuming or impossible, while physical interference with nets makes fishing more difficult or impractical. Some fishing communities in West Africa have been abandoned as a direct result of the arrival of water hyacinth.

Water hyacinth has direct effects upon water chemistry. It can absorb large amounts of nitrogen and phosphorus, other nutrients and elements. It is this ability to pick up heavy metals which has led to the suggestion that water hyacinth could be used to help clean industrial effluent in water. By absorbing and using nutrients, water hyacinth deprives phytoplankton of them. This leads to reduced phytoplankton, zooplankton and fish stocks. Conversely, as the large amounts of organic material produced from senescent water hyacinth decompose, this leads to oxygen deficiency and anaerobic conditions under the floating water hyacinth mats. These anaerobic conditions have been the direct cause of fish death, and changes in the fish community by eliminating most species at the expense of air breathing species. Stationary mats of water hyacinth also shade out bottom growing vegetation, thereby depriving some species of fish of food and spawning grounds. The potential impact on fish diversity is enormous. The conditions created by water hyacinth encourage the vectors of several human diseases, including the intermediate snail hosts of bilharzia (schistosomiasis) and most mosquito vectors, including those responsible for transmission of malaria, encephalitis and filariasis. In parts of Africa, water hyacinth mats are reported to provide cover for lurking crocodiles and snakes.

The diversity of impact means that the problems occur in the mandates of diverse ministries. There is considerable scope for delays following a new infestation while the relevant government groups decide who is responsible for what in order to tackle water hyacinth.

Prepared by Matthew Cock, CABI Bioscience Switzerland Centre, 1 Rue des Grillons, CH-2800 Delémont, Switzerland. www.cabi.org/bioscience/switz.htm.

CASE STUDY 5.2 Paper-bark Tree Alters Habitats in Florida

Melaleuca quinquenervia, the paper-bark tree, is an evergreen tree with a slender crown, which grows up to 29 m tall. It has white many-layered papery bark and white flowers in brush-like spikes. It is native to Australia and Papua New Guinea and was introduced to Florida at the beginning of the 20th century to provide a useful crop that would grow in an area subject to drought, flooding and periodic fires where little else was productive. Although hopes of using paper-bark tree for timber were not fulfilled, it did prove economical to produce as an ornamental.

But it was an unfortunate choice for an introduction. It grows phenomenally fast in Florida (18-month-old trees can be 6-7 m tall) and it flowers up to five times a year. Its wind- and water-dispersed seeds are produced from trees as young as two years old, and are retained on the tree to be released in times of stress - fire, frost and herbicide all cause seed capsules to open. Mature tree can hold up to 20 million seeds; on the tree they can remain viable for up to 10 years, but viability is lost quickly once the seeds are in the soil. *M. quinquenervia* grows densely, forming impenetrable thickets, and also spreads by adventitious roots, which cause soil accretion to occur owing to thick mats of roots at the water surface, and this leads to an increase in the elevation of the infested area. Small increases in elevation of a few centimetres make huge differences in the composition of Everglades plant communities, so *M. quinquenervia* is converting wetland to upland in this manner. It is adapted to subtropical climates with a preference for seasonally wet sites and flourishes in standing water. In the last 30-40 years it has spread rapidly and now infests close to half a million acres (some 200,000 ha) in south Florida, causing extensive environmental and economic damage particularly in the Everglades where it threatens the native habitat.

Edited from a <u>Biocontrol News and Information</u> news item by Dr Gary R. Buckingham, USDA/ARS, Biocontrol of Weeds, c/o Florida Biocontrol Laboratory, P.O. Box 147100, Gainesville, FL 32614 7100, USA. E-mail: <u>grbuck@nervm.nerdc.ufl.edu</u>

CASE STUDY 5.3 Chestnut Blight Changes a Forest Ecosystem

The demise of the American chestnut (*Castanea dentata*) illustrates how an entire ecosystem can be fundamentally altered. Until early in the 20th Century, the chestnut was one of the most abundant hardwoods of the eastern deciduous forests of the USA, in some areas accounting for as much as 25 percent of all trees. It was also among the most economically important trees in the eastern USA, with wood that was highly valued for furniture and construction, and nuts that were both a cash crop and a staple for wildlife. In the early 1900s, a fungal chestnut blight (*Endothia parasitica*) from China was introduced accidentally, killing as many as one billion trees over 91 million acres. Although the American chestnut still survives as a species, it is ecologically extinct - no longer a functional part of the ecosystem. Its loss has permanently changed the ecology of the eastern deciduous forests.

Edited from: Stein, Bruce A. and Stephanie R. Flack, eds. 1996. America's Least Wanted: Alien Species Invasions of U.S. Ecosystems. The Nature Conservancy, Arlington, Virginia. Available through <u>http://www.tnc.org</u>/

CASE STUDY 5.4   Hybridisation

Mating between some introduced and native species can lead to extinction of the native species by replacing some of its genes. For example, mallards (*Anas platyrhynchos*) introduced to the Hawaiian Islands for hunting have hybridised extensively with the native, endangered Hawaiian duck, greatly complicating recovery plans for the latter species. On the USA mainland, mallards migrate to Florida in the winter.   Although they formerly bred only while in the North, domesticated mallards released to the wild in Florida for hunting have bred with the native Florida mottled duck (*Anas f. fulvigula*), whose existence may now be threatened by hybridisation. A similarly critical situation arose from the introduction of mallards into New Zealand, where they hybridise with the endemic subspecies of the grey duck (*Anas s. superciliosa*).

Rainbow trout (*Oncorhynchus mykiss*) introduced into western USA watersheds as sport fish hybridise extensively with the Gila trout and the Apache trout - two species that are listed under the Endangered Species Act.

Plants can also fall prey to the same insidious phenomenon. An example is *Lantana depressa*, which is found on a few dune and limestone ridge habitats of peninsular Florida. It hybridises with *Lantana camara*, the descendant of several Latin American or West Indian species that were brought to Europe as ornamentals in the seventeenth century, hybridised by horticulturists, and then introduced by the late 18th century into the New World.

Even if no genes are exchanged between the hybridising species, the process can threaten the existence of one of them. Introduced brook trout (*Salvelinus fontinalis*) are today displacing native bull trout (*S. confluentus*) in parts of the western USA. Although there is extensive hybridisation between the two species, the hybrid offspring are sterile, so they cannot transmit brook trout genes back into the bull trout population. But the loss of productive mating opportunities by the rarer species, the bull trout, contributes to its displacement.

At least three of the twenty-four known "extinctions" of species listed under the U.S. Endangered Species Act have been wholly or partially caused by hybridisation, and there seems to be no limit to other possible consequences. Johnson grass (*Sorghum halepense*) was originally introduced to the USA around 1800 as a forage crop for cattle and is now viewed as one of the worst weeds. Among other noxious traits, it hybridises with cultivated sorghum to produce "shattercane," which is agriculturally worthless.

Even worse outcomes are possible. For example, North American smooth cord-grass (*Spartina alternifolia*), which entered England in the holds of ships in ballast soil, hybridised there with an innocuous native species (*S. maritima*) to produce new plants (*S. x townsendii*) that proved sterile. This might have ended the story, had not one of them undergone a chromosome doubling, yielding a new species that turned out to be a fertile, invasive weed (*S. anglica*).

Edited from Simberloff, D. (1996) Impacts of Introduced Species in the United States Consequences 2(2), 13-23.

CASE STUDY 5.5   Eradication of a Deliberately Introduced Plant Found to be Invasive

Kochia, *Bassia scoparia*, was introduced into Western Australia in 1990 and promoted as a 'living haystack'. It was widely planted on 52 properties in the south-west of the State. Its rapid growth, suitability as forage and high salt tolerance promised farmers new uses for their marginal salt land and the seed was added into general salt land seed mixes.

Early in 1991 one farmer noticed shrubs growing alarmingly well within, and spreading from, his saltland rehabilitation plantings. He was worried enough to call Agriculture Western Australia (AWA) and voice his concern about this 'kochia' plant. He later ploughed in the whole site destroying all plants before seed was set. An agency researcher then visited another site, confirmed the identification, and raised the alarm with the Weed Science Group of AWA.

A literature search revealed hundreds of articles on the impact and invasiveness of kochia and the Weed Science Group spent the next few months determining control and management options and documenting and surveying all sites where kochia had been planted. Two pamphlets, or 'Farmnotes', were produced and extensive use made of radio, television and the print media to alert farmers to the problem. Kochia was gazetted as a 'Declared Plant' early in 1992 and the eradication campaign started.

The rapid spread of kochia was alarming, from 52 properties in 1991 to over 270 two years later. Large plants where found to have tumbled over 5 kilometres from their point of origin and seen to roll over fences! The next year hundreds of seedlings appeared along the line of the previous years tumble tracks. The extent of the infestations was huge with the most northerly infestation over 900 km from the most southerly. Staff and resources were spread thin.

Over the next eight years over 21,345 ha of property was searched and 4989 ha treated or programmed for treatment by AWA field staff. At its peak in 1993 the total area infested was determined to be 3,277 ha. By 1995 that area was reduced to 139 ha and by 2000 two properties were programmed for treatment, for a total of 5 ha.

Eradication is considered successful at a site following three clean years of inspection. The vast majority of properties have been clean for the past three years. Only four properties out of 270 are yet to be considered clean; this amounts to an almost 99% success rate for the programme to date. Total costs have been estimated at Australian $500,000 over the eight years from 1992 to 2000.

The essential keys to success are considered to have been:

➤ Rapid response to an identified threat. Weed Science staff were working on eradication just a few months after the plants were found.

➤ Excellent surveillance. Extensive use of field staff with their local knowledge and the media was used to determine where to start looking for plants.

➤ Exemplary landholder co-operation. The very last infested property found was reported by a landholder and landowners were very generous with their time, resources and knowledge in assisting agency staff throughout the campaign.

Prepared by Rod Randall, Weed Risk Assessment, Weed Science Group, Agriculture Western Australia http://www.agric.wa.gov.au/progserv/plants/weeds/

CASE STUDY 5.6   Eradication Programme for Chromolaena Weed in Australia

Chromolaena or Siam weed (*Chromolaena odorata*) was discovered in Australia in 1994, (see Case Study 4.5 "Detection of Chromolaena Weed in Australia"). It was declared a noxious weed several years earlier, enabled a rapid survey and eradication campaign to be mounted (see Case Study 5.12 "Surveying for Chromolaena Weed Infestations in Australia").   The Queensland Department of Natural Resources (DNR) is working with other government departments and the community to eradicate this weed in a five-year eradication programme with an annual budget of around A$170,000.

Registration of two chemicals for use in control of *C. odorata* was "fast-tracked" immediately after its discovery (one as an overall spray, and one for basal bark treatment), and chemical control at the correct rates gives excellent weed kill.   Intensive weed management practices on sugarcane and banana plantations along the Tully River have also probably kept the infestation in check in these areas.

A mix of control techniques has evolved over the five-year period in response to experience and field testing. Changes have been needed to cope with factors such as

➤   while the most common plant flowers May to July (thought to be triggered by the shortening daylength), a second phenotype flowers during March

➤   the unusual double-flowering discovered in 1998/99

➤   when the seasons are less pronounced, flowering can be erratic

➤   the seeds appear to be viable for longer than the four years originally thought

➤   viability of seeds is achieved sooner in flower development than was first thought

The control techniques used have also evolved over time to cope with the changing nature of the campaign as plant numbers at old sites are reduced and more recent discoveries extend the distance to be covered in each control operation. Plants are more difficult to find at old locations and accurate mapping, follow up programmes and good local knowledge therefore become more important.   More time is being spent finding fewer plants, and even under the most favourable conditions, plants will still be missed on occasion.   Consistent follow-up work is therefore absolutely critical to the success of this eradication campaign.

Competing vegetation quickly establishes itself at sites where the larger *C. odorata* plants have been removed and exotic grasses can mask existing plants and also hinder germination of *C. odorata* seeds. Field trials using glyphosate to kill these grasses and leave the weed seed free from competition have proven effective and have become an important tool to facilitate germination so that the seedlings can be killed.

Plant numbers are so reduced along Echo Creek (site of the heaviest original infestations) that it is now possible to walk the majority of its length hand pulling any Chromolaena encountered. Occasional hot spots can be followed up later. Along the Tully River only a few of the original sites had any seedlings left and numbers at these were very low.

At the end of five years, the populations of Chromolaena have been dramatically reduced, but not yet eradicated. With the exception of a single plant (75 km inland), all other infestations are within a 50 km radius of the original sighting at Bingil Bay.

Edited from: http://www.dnr.qld.gov.au/resourcenet/fact_sheets/pdf_files/pp49.pdf, the DNR Pest Facts web-page on Siam Weed and unpublished DNR reports.

Three hundred years ago Phillip Island must have been mostly covered by sub-tropical rainforest, like its neighbour, Norfolk Island, in the South Pacific, half way between Australia and Fiji. The uninhabited, 260ha island is very rugged, with cliffs up to 250m and some areas inaccessible by normal means. Pigs were introduced about 1790, and goats and rabbits shortly after. The vegetation was soon severely damaged. Pigs had died out or been killed by about 1850, and the goats survived until about 1900. The rabbits remained. A programme, started as an experimental exercise to demonstrate the effects of rabbits on Phillip Island, proceeded to their complete eradication.

In 1978 the island was mostly devoid of vegetation although some patches remained. Most of the ground was bare and heavily eroded. Erosion was continuing and the surrounding sea turned brown after heavy rain. A series of experimental exclosures to demonstrate the effect of grazing by the rabbits were set up. Although the experimental programme was intended to be for three years, the results were so dramatic that after the first year a decision was made to eradicate the rabbits. Sites, which were devoid of vegetation, became thickly vegetated (mainly by weeds) when fenced off from the rabbits. In some exclosures 22 plant species were identified growing within six months.

The first approach to eradicating the rabbits was by using a very virulent strain of myxoma virus, using European rabbit fleas as the vector. Disease-free fleas were first introduced, and then two months later fleas carrying myxoma were delivered. The island is so rugged that ropes and climbing techniques were used to reach some parts, other parts were reached by swimming from a boat 150m offshore, and vials of fleas were shot by bow and arrow to places that could not be reached by other methods.

The decline in the rabbits was dramatic, and vegetation started to appear on the bare ground. At that stage, seedlings of *Abutilon julianae* were found. This species had not been recorded from Phillip Island before and was last seen on Norfolk Island around 1910. It had been believed extinct, but must have survived in some part of Phillip Island inaccessible even to rabbits. Introductions of myxoma-carrying fleas were continued, because the disease was too virulent for natural transmission through the remaining rabbits. Unfortunately the supply of fleas from Australia stopped and the rabbits began increasing again. Consequently other methods were used to kill the remaining rabbits. In 1983, 350 bait stations were established and the rabbits were poisoned with "1080". Trapping, gassing and shooting killed the last surviving rabbits. The last rabbit was shot on an inaccessible ledge in 1988.

The whole exercise was labour-intensive, but Phillip Island has considerable value for nature conservation, and importantly, the island does not have rats, mice or cats. It is used by seabirds for breeding, has an endemic hibiscus (*Hibiscus insularis*), which was reduced to just a few plants at two sites, and has some endemic invertebrates (including a centipede, *Cormocephalus coynei*, and a cricket, *Nesitathra philipensis*).

Prepared by Peter Coyne, Environment Australia, Canberra, Australia,
http://www.biodiversity.environment.gov.au/protecte/alps/

CASE STUDY 5.8   Eradication of the Giant African Snail in Florida

The giant African snail, *Achatina fulica*, about three inches long, has been introduced widely in Asia, to islands in the Pacific and Indian Oceans, and recently to the West Indies. It is seen as a serious agricultural pest, and predatory snails such as *Euglandina rosea* that were introduced to attack it have only added to the problem by extinguishing many native snail species (see Case Study 3.1 "Rosy Wolfsnail, *Euglandina rosea*, Exterminates Endemic Island Snails"). It was, however, successfully eradicated from Florida - although neither easily nor cheaply.

In 1966, a boy returning from Hawaii smuggled three of the snails into Miami, and his grandmother released them in her garden. Reproduction ensued, and in 1969 the Florida Division of Plant Industry (DPI) was alerted, leading to an immediate survey. The state Commissioner of Agriculture notified the news media about the giant snail, mailed over 150,000 copies of an attractive brochure, and called for public assistance in reporting and eliminating it. An area covering about forty-two city blocks was quarantined, but within days, a second infestation was discovered - in Hollywood, 25 miles north of Miami and well outside the initial quarantine zone.

The ensuing eradication campaign relied primarily on hand-picking, plus a granulated chemical bait. There were frequent surveys, and by 1971 in a six months period only forty-six snails were found - compared to 17,000 in the previous sixteen months. In Hollywood, seventeen months after its initial infestation, only one adult snail was found. But less than a month after the effort seemed to have succeeded, a third infestation, probably three years old, was discovered three miles south of the original Miami site, with over 1,000 live snails on one block. The block was quarantined, and a large buffer zone was surveyed and treated. Nine months later, a fourth infestation, again about three years old, was found two miles north of the original one, followed by a fifth, about half a mile north of the initial infestation.

Although profoundly disappointed, the DPI persisted. By 1973, seven years after the three snails were brought into the city, more than 18,000 had been found, and many eggs. In the first half of that year, by contrast, only three snails were collected, in two sites. By April of 1975, no live specimens had been found for almost two years, and the campaign - which had cost over US $1,000,000 - was judged successful. Frequent surveys were continued for many months, along with the application of bait and chemical drenching. As a result, the giant African snail has not been found again, anywhere in the state.

Edited from Simberloff, D. (1996) Impacts of Introduced Species in the United States. Consequences 2(2), 13-23.

CASE STUDY 5.9   Eradicating Screwworms from North America and North Africa

Screwworms, the larvae of the screwworm fly, are parasites that cause great damage by entering open wounds and feeding on the flesh of livestock and other warm-blooded animals, including humans. The New World screwworm fly (*Cochliomyia hominivorax*) is native to the tropical and sub-tropical areas of North, South, and Central America, while similar but less damaging species occur in the Old World.

After mating, the female screwworm fly lays her eggs in open wounds. One female fly can lay up to 400 eggs at a time, and as many as 2,800 eggs during its lifespan of about 31 days. The screwworm grows to more than 1 cm inside the wound within a week of entering the wound. The full-grown larva then drops from the wound, tunnels into the soil, and pupates before emerging as an adult screwworm fly. Left untreated, screwworm-infested wounds lead to death. Multiple infestations can kill a grown steer in 5-7 days. Losses to livestock producers in the USA had exceeded $400 million annually.

The sterile insect technique (SIT). Screwworms are eradicated through a form of biological control. Millions of sterile screwworm flies are raised in a production plant located in the southern Mexican State of Chiapas. During the pupal stage of the fly's life cycle, the pupae are subjected to gamma radiation. The level of radiation is designed to leave the fly perfectly normal in all respects but one: it will be sexually sterile. Thus, when the artificially raised flies are released into the wild to mate with native fly populations, no offspring will result from the matings. These unsuccessful matings lead to the gradual reduction of native fly populations. With fewer fertile mates available in each succeeding generation, the fly, in essence, breeds itself out of existence.

In the early 1950s, USDA's Agricultural Research Service developed the SIT for screwworm control. This SIT was used operationally in Florida in 1957, and by 1959, screwworms had been eradicated from the Southeast USA. The SIT was next applied in the more extensively infested Southwest starting in 1962. Self-sustaining screwworm populations were eliminated from the United States by 1966. Since then, a co-operative international programme has been pushing the screwworm back towards the Isthmus of Panama, with a view to eradicating it from Central America, and in the future, the Caribbean.

Hence, when an infestation of the New World screwworm appeared in Libya in 1988, the tools for its eradication were already available. Recognising the enormous threat to humans, livestock and wildlife, an urgent national and international effort was mounted to prevent its spread to the rest of Africa and the Mediterranean Basin. The SIT campaign was successful in achieving eradication, preventing the enormous losses that would have occurred if the infestation had spread.

Edited from the USDA-APHIS web-site http://www.aphis.usda.gov/oa/screwworm.html.

CASE STUDY 5.10 Fire Ant: an Eradication Programme that Failed

Trying to eradicate every single individual of a harmful introduced species is a seductive but controversial goal. One would like to eliminate ongoing, and sometimes increasing, damage, but the development of eradication technology may prove daunting, and a failed attempt may be exceedingly costly and invoke colossal damage to non-target species.

For example, the attempt to eradicate the introduced fire ant (*Solenopsis invicta*) from southern states of the USA proved disastrous. In 1957, a well-meaning Congress authorized US $2.4 million for the project, but the initial heptachlor applications caused wildlife and cattle deaths. Researchers next developed mirex bait, but the introduced fire ant rapidly re-invaded areas from which it was eliminated much more quickly than indigenous ant species, thereby increasing the populations of fire ant due to lower competition. Additionally, mirex residues were discovered in many non-target organisms. These findings resulted in a ban by the Department of Interior on use of the insecticide on its lands. Registration of mirex was finally cancelled by the Environmental Protection Agency in 1977, but by then the costs of the applications had climbed to about US $200 million and the range of fire ants had only expanded, dramatically, during the eradication campaign.

Edited from Simberloff, D. (1996) Impacts of Introduced Species in the United States Consequences 2(2), 13-23.

CASE STUDY 5.11 Colonization Rate of Hibiscus Mealybug in the Caribbean

The pink hibiscus mealybug (*Maconellicoccus hirsutus*) occurs in most tropical areas of the world, including Asia, the Middle East, Africa, Australia and Oceania. It arrived on the Caribbean island of Grenada in the early 1990s but it was not until November 1994 that it was officially reported. It has a very wide host range, and rapidly became a very serious plant pest attacking more than 200 genera of plants. It is particularly associated with species of Malvaceae such as ornamental hibiscus, and the indigenous watershed tree, blue mahoe (*Hibiscus elatus*), but also caused severe damage to a wide variety of indigenous and exotic plants including fruit trees, samaan trees, annonas, cacao, teak, etc. It was a national disaster. In addition to the direct damage caused, the mealybug rapidly disrupted trade, as nearby islands started to ban trade in fruit and vegetable produce from Grenada.

One of Grenada's major trading partners to whom it exported much fruit, vegetable and flower produce is the nearby island of Trinidad. In 1995, interceptions at Trinidad & Tobago's Port of Spain docks were soon being reported and produce had to be dumped at sea. For a while Trinidad & Tobago attempted to prevent the entry of the hibiscus mealybug from Grenada, but the volume and diversity of boat and plane traffic made this a battle that the Ministry of Agriculture could not win. By August 1995, it became clear that half a dozen or more separate introductions had occurred in different parts of the country, and the Ministry moved on to a containment programme, before in due course largely solving the problem through a successful biological control programme.

The original introduction into Grenada was a freak event, and the actual pathway of introduction is unknown. However, once the mealybug became established in Grenada in huge populations on a wide variety of plants in urban and rural situations, the opportunities for spread to neighbouring islands became very common. Under these conditions, its spread through the region became inevitable. Reports of mealybug invasions came in from several Caribbean islands over the next few years, and by 1998 it had been reported from over 15 territories from Guyana in the south to Puerto Rico in the north. There is no doubt that the successful implementation of a biological control programme subsequently slowed the rate of spread substantially. By decreasing the established populations to a relatively low level, the frequency of contamination of fruit and other articles moved between islands also went down, and there are still areas in the Caribbean yet to be colonized.

Prepared by Matthew Cock, CABI Bioscience Switzerland Centre, 1 Rue des Grillons, CH-2800 Delémont, Switzerland. www.cabi.org/bioscience/switz.htm.

CASE STUDY 5.12 Surveying for Chromolaena Weed Infestations in Australia

Chromolaena weed (*Chromolaena odorata*) was discovered in Queensland, Australia, in 1994 (see Case Study 4.5 "Detection of Chromolaena Weed in Australia").

The Department of Natural Resources (DNR) is responsible for weeds legislation and control in Queensland. Pre-emptive declaration of *C. odorata* as a noxious weed several years prior to its discovery in Australia enabled a rapid survey and eradication campaign to be mounted. In all shires of Queensland it is classed as:

➤ Category P1 - introduction into Queensland is prohibited

➤ Category P2 - where discovered, it is to be destroyed

All sightings must be reported to allow immediate control by Department of Natural Resources staff. Other Australian states where *C. odorata* poses a threat have also declared it a noxious weed.

Surveys prior to control programmes provide critical information on precise location and probable densities of infestations. Surveys are conducted only by DNR staff, but can be related to information received from local people or from State/Local Government agencies.

Helicopter surveying has proven to be a cost-effective solution in areas where access is a problem. In certain instances it is the only method available and is extremely useful when time is short, usually towards the end of control programmes. As the main control programme is concluding at the time of flower development, this is found to be the ideal time for aerial observations i.e. when the plant is most visible. Flying is done at treetop level and at less than 10 knots ground speed. Quite small plants have been found in this way and this method can be used as a follow-up technique in some instances. Plant locations are fixed by:

➤ dropping numbered streamers onto the ground,

➤ marking positions on aerial photographs carried in the aircraft, and

➤ recording GPS positions.

This method has proved 100% reliable and experience has shown that all three practices are necessary to eliminate frustrating delays on the ground in following up.

Ground surveying continues to be undertaken by DNR control staff concentrating mainly on areas adjacent to any recent discoveries. Extension initiatives by these staff are aimed at increasing local awareness and fostering a spirit of co-operation among landholders with whom a rapport has been achieved. Awareness has been raised amongst Local Government staff who regularly work on roadsides and easements, etc. and valuable contributions have since been made by many of these.

Edited from: http://www.dnr.qld.gov.au/resourcenet/fact_sheets/pdf_files/pp49.pdf, the DNR Pest Facts web-page on Siam Weed and unpublished DNR reports.

CASE STUDY 5.13   Controversy Over Mammal Control Programmes

Controversies over the management of feral horses in both the USA and New Zealand illustrate the conflicts that readily arise between environmentalists and other segments of society about some widely appreciated feral domestic animals. In both countries feral horses pose documented threats to native species and ecosystems. Yet some groups contend the horses that escaped from Spanish explorers in North America about 500 years ago "belong" in the West, merely serving as replacements for native equids that became extinct on the continent about 10,000 years ago. In New Zealand, however, there were no native land mammals, except for bats, before humans arrived some 900 years ago. European settlers introduced horses into New Zealand less than 200 years ago.

In New Zealand, feral horses have occupied the central North Island since the 1870s. Land development and hunting progressively reduced their numbers to about 174 animals in 1979. By 1981, however, public lobbying resulted in creation of a protected area for the remaining horses. With protection, horses increased to 1,576 animals by 1994, essentially doubling their population every four years. In response to damage in native ecosystems caused by this rapidly growing population, the New Zealand Department of Conservation recommended management to retain a herd of about 500 animals. The management plan, which included shooting horses, provoked intense public protest. This outcry eventually resulted in the overturning of a scientifically based management plan and a 1997 decision to round up as many horses as possible for sale. Sale of several hundred horses duly took place, but the long-term fate of the growing herd remains unresolved. The impasse in New Zealand over feral horse control has been mirrored in Nevada, where an intense dispute has raged between land managers and pro-horse activists about the ecological impacts of feral horses, the size of feral herds, and appropriate methods of population control. At a practical level, the removal of animals by culling would probably be the simplest way of achieving population reduction, but public resistance precludes this option.

The infusion of strong public sentiment into policy for feral horses, as well as burros in the U.S., would likely serve as a mild preview of public reaction to serious efforts to control feral cats. Ample evidence demonstrates that feral cats are the most serious threat to the persistence of many small vertebrates. One study in Britain estimates that domestic cats alone kill 20 million birds annually; the toll for feral cats, while unknown, clearly adds to this tally. The degree to which feral cats in Australia should be eradicated and domestic cats sterilized has already engendered vituperative debate. Similar discussion, pitting environmentalists against the general public, is being played out in the USA and Europe. Few biotic invasions in coming decades will deserve more even-handed comment from ecologists than the dilemma caused by feral cats.

Edited from: Mack, R.N.; Simberloff, D.; Lonsdale, W.M.; Evans, G.; Clout, M.; Bazzaz, F. (2000) Biotic Invasions: Causes, Epidemiology, Global Consequences and Control. Issues in Ecology Number 5, 22 pp. (http://esa.sdsc.edu/issues5.pdf)

CASE STUDY 5.14   Containment of the Spread of Chromolaena Weed in Australia

Chromolaena or Siam weed (*Chromolaena odorata*) was discovered in Queensland, Australia in 1994, (see Case Study: 4.5 "Detection of Chromolaena Weed in Australia"). The primary (i.e. original) infestation at Tully River, and various secondary infestations were located. River-borne spread of Siam weed away from the primary infestation was evident. Some secondary infestations were probably initiated from seed carried away from the Tully River by other means. It is crucial that chromolaena weed is not allowed to spread further outside the current infested areas while an eradication programme is carried out (see Case Study 5.6 "Eradication Programme for Chromolaena Weed in Australia"). Potential pathways for spread include:

➤   Tully River sand used unsterilised in plant and palm potting mixtures

➤   Movement of equipment, e.g. earthmoving operations

➤   Movement of farm stock

➤   Pasture seed sold from the Mission Beach area

➤   Bush walking and cross country sports

➤   Backpackers camping in infested areas

➤   Livestock (cattle and horses) and feral pigs

➤   Mechanical clearing along power line routes

Department of Natural Resources (DNR) extension officers provide inputs with the aims:

➤   That all levels of government and the general public understand the importance of the weed and the need for eradication. This includes selling and distributing the message of excluding this invasive plant from the sparsely populated areas of Cape York Peninsula where detection and control problems are compounded by the isolation of the area.

➤   To have sufficient numbers of people who can identify the plant to ensure a rapid reaction to any suspect sighting.

➤   To provide encouragement for continued reporting of any suspected sightings.

Recent activities have included:

➤   TV coverage of a new infestation

➤   Individual contact with all surrounding neighbours of all newly found infestations

➤   Potted specimens taken to agricultural field days and agricultural shows

➤   Identification competitions at these shows and field days. People pick out Siam weed from its local 'look-alikes', encouraging activity learning of the identifying features.

➤   Presentations on "Problems and Identification of Siam Weed", given to community groups (e.g. Landcare, Cane Growers, Aboriginal communities etc.)

These efforts have increased local awareness and fostered a spirit of co-operation among landholders with whom a rapport has been achieved. Awareness has also been raised amongst Local Government staff who regularly work on roadsides and easements, etc. and valuable contributions have since been made by many of these.

Edited from: http://www.dnr.qld.gov.au/resourcenet/fact_sheets/pdf_files/pp49.pdf, the DNR Pest Facts web-page on Siam Weed and unpublished DNR reports.

CASE STUDY 5.15   Containment vs. Eradication: *Miconia calvescens* in Hawai'i

In formulating strategies for combating *Miconia calvescens* in Hawai'i, it has been difficult to settle on the goal to be achieved - whether it is eradication or containment. Because of the life history characteristics of *M. calvescens*, with 45 years and 3 m of height growth separating seedlings from fruiting trees, eradication is clearly possible for some small localized populations and may be possible for entire islands. On the other hand, the longevity of the seed bank (3+ years, but not fully assessed) means that an eradication programme has to be continued for several years to ensure success. Eradication will clearly require sustained commitment and funding for many years, but so will containment. The control programme in Hawai'i aims for eradication at a local or island level, realizing that complete eradication of the weed may not be achieved. Unfortunately, as long as there is a seed source of *M. calvescens* in the state, there will always be a strong possibility of infecting new areas and re-infecting areas from which the plant has already been eradicated.

Edited from "Miconia calvescens in Hawaii: a summary" prepared by L. Loope (March 1996), with extensive borrowings from manuscripts by Medeiros, Loope and Conant and by Conant, Medeiros and Loope, and posted on the internet at http://www.hear.org/miconiainhawaii/miconiasummarybylll.htm.

CASE STUDY 5.16   Seed Movement on Vehicles: a Study from Kakadu National Park, Australia

In order to investigate the importance of cars as vector of weed dispersal, seeds were collected at roughly monthly intervals, during May 1989 to May 1990, from tourist vehicles parked overnight during a two day period at a campsite in Kakadu National Park, northern Australia. A total of 1960 seeds were collected from 304 tourist vehicles by vacuuming the radiator and outer surfaces of the car and by sampling mud from the wheel arches and tyres. Individual cars were found to carry up to 789 seeds and a maximum of 15 species, but the majority (96%) of cars carried one or no seeds. The proportion of cars carrying seeds, and the total number of seeds entering per month, did not vary strongly with seasons, despite the fall in numbers of cars entering during the wet season.

The numbers of seeds and occurrences of different weed species on the cars were unrelated to the abundance of the weeds found previously in the park. However, those weed species that were found on tourist cars occurred at three times as many sites in the park as those that were not, suggesting that movement of seeds by tourist cars may be partly responsible for weed infestations.

Most (66%) of the 88 species in the samples were grasses. Ten species of known tropical weeds were found amongst the samples, including *Pennisetum polystachion*, *Sida acuta*, *Hyptis suaveolens*, *Cenchrus ciliaris* and *Tridax procumbens*, as well as 14 species not known in the park. Propagules of the major invasive tropical weeds, *Mimosa pigra* and *Salvinia molesta*, were not found amongst the samples.

It is concluded that, in view of the low density of weed seeds entering the park on tourist vehicles, resources are best spent on detecting and eradicating existing weed infestations, rather than on attempting to prevent this form of seed movement.

Abstracted from Lonsdale, W.M.; Lane, A.M. (1994) Tourist vehicles as vectors of weed seeds in Kakadu National Park, Northern Australia. Biological Conservation 69, 277-283.

CASE STUDY 5.17 Reptile Recovery on Round Island

Round Island, near Mauritius, has been a refuge for rare and endangered reptiles, plants, seabirds and invertebrates, many found nowhere else in the world. These species once had a wider distribution, including the main island of Mauritius. Eight or so thousand years ago, rising sea levels marooned the species on Round Island and, subsequently, the mainland populations died out after rats, cats and other alien animals were introduced by human colonists. Round Island provided a refuge – somehow these species survived into the 20th century on this small island which was itself rapidly disappearing. Introduced rabbits and goats were eating all of its vegetation, and the soil was slipping into the sea.

Some representatives of three of the rarest reptiles – Guenther's gecko, Telfair's skink and Round Island boa – were brought to Jersey Zoo to start a captive-breeding programme, and efforts were consolidated to remove rabbits and goats from the island and halt the erosion.

Rats, cats and mice have been eradicated from several small islands around Mauritius and Rodrigues. Slowly, these islands are being restored to a state where they can once again support communities of native reptiles. In addition to the three Round Island reptiles, there are several other endangered skinks and geckos that will benefit from these efforts.

The Guenther's gecko, one of the largest and rarest geckos in the world, has not increased markedly in number on Round Island since the goats and rabbits were removed. Its population is still only in the hundreds of individuals. They feed on insects and nectar, but it has recently been discovered that they can also be effective predators of small geckos. This may preclude translocating them to islands that have populations of other rare species of gecko. A suitable island will be chosen on which they can be introduced to establish an additional population to reduce the risk of extinction by freak events.

In contrast, Telfair's skink has increased dramatically on Round Island, with a population of several tens of thousands. However, it must still be considered vulnerable since it is found only on the one island. Great caution would have to be exercised in any future decision to translocate Telfair's skinks. They are very predatory and might otherwise end up feeding on some other highly endangered species of reptile.

Edited from: Carl Jones, Mauritius Programme Director, in On the Edge No. 83 (November 1998).

CASE STUDY 5.18   Conservation Management Areas in Mauritius

Intensively managed vegetation plots have been established in representative vegetation communities of Mauritius to conserve plant genetic resources. The first plot was established in the upland forest of Macchabee in the 1930s by Dr. Vaughan, the then Conservator of Forests. There are now eight extensively managed plots, Conservation Management Areas (CMAs) as they are called, ranging from 1.5 ha to 19 ha within the National Park. These CMAs are fenced and a low stone wall was built to keep deer (*Cervus timorensis*) and pigs (*Sus scrofa*) out and the weeds are manually uprooted.

The fencing and initial weeding of most of the CMAs and the maintenance weeding four times a year in all the eight CMAs, covering an area of 38 ha has been contracted out because of shortage of manual labour within the National Parks and Conservation Service. The exotics being removed from within the CMAs include *Ardisia crenata, Camellia sinensis, Clidemia hirta, Desmanthus virgatus, Eucalyptus* spp., *Eupatorium pallescence, Homalanthus populifolius, Lantana* spp., *Ligustrum robustum, Litsea* spp., *Mimosa pudica, Pinus* spp., *Psidium cattleianum, Ravenala madagascariensis, Rubus alceifolius, R. roseifolius, Stachytarpheta jamaicensis, Syzygium jambos,* and *Wikstroemia indica.*

Volunteers from Raleigh International have tried some chemical control within the now extended Brise Fer CMA during six weeks in 1993. Chinese guava (*Psidium cattleianum*) and privet (*Ligustrum robustum*) were cut with machetes at about waist height and herbicide was applied to the stump with a small brush at a concentration of 10% (one part garlon to 9 parts water) and a few drops of rhodamine dye were added for identification purposes. The conditions were generally moist during this period and not ideal for application of garlon. Other attempts at control of the two invasive plants by using garlon at a higher concentration of about 20% did not produce promising results as the herbicide only retarded the formation of new shoots.

The control of the invasive alien plant species in these CMAs has proved to be very promising. Many endangered plants have been found in the CMAs, the endemics are regenerating naturally and they are providing better habitat for the endemic birds. Only two known specimens of *Claoxylon linostachys* were known from Macchabee before the establishment of the plot at Mare Longue where a population of about 20 individuals has been discovered. The CMAs are being used by the endemic pink pigeon (*Nesoenas mayeri*) and the echo parakeet (*Psittacula echo*) for nesting and foraging.

Edited from "Control of alien invasive species and exotic fauna", a paper presented at the Global Invasive Species Programme workshop on Management and Early Warning Systems, Kuala Lumpur, Malaysia, 22-27 March 1999, by Dr. Yousoof Mungroo, Director, National Parks and Conservation Service, Ministry of Agriculture, Fisheries and Co-operatives, Reduit, Mauritius.

CASE STUDY 5.19   Mechanical and Chemical Control of Seastars in Australia are Not Promising

The Northern Pacific seastar (*Asterias amurensis*), common in the seas around Russia and Japan, and extending south to Korea and east to Alaska, was probably introduced to Australia in the early 1980s, and was recorded until last year only in the Derwent estuary, site of Tasmania's major port. Its spread from the Derwent is believed to have been restricted by the estuarine circulation, but it has recently been found in Victoria's major port, Prince Phillip Bay. Genetic tests indicate that this introduction was most likely through the Tasmanian population, the probable vector - shipping. The seastar feeds voraciously and omnivorously on shellfish, and virtually all sizeable bivalves and other attached or sedentary invertebrates are eliminated where seastar densities are high. This may not only affect the biodiversity, but also have effects on the ecosystems of which bivalve filter-feeders are a key component. The seastar threatens the fisheries of southern Australia - even in its native range it has a significant impact on fish and shellfish productivity. It has wide temperature and salinity tolerances, and populations in the Derwent estuary have grown to the point where it is the dominant invertebrate predator of some benthic communities. Population densities easily exceed those recorded in its native range - one estimate puts the Derwent Estuary population at 30 million. Its impact is considered so significant that ports in Tasmania, and now Port Phillip Bay, are the only ports in the world from which ships are prevented from releasing ballast water in New Zealand's coastal waters under any conditions.

Physical removal of seastars using divers or traps has been tested. Community divers removed 30,000 seastars from around the Hobart wharves on two occasions in 1993, or perhaps 60% of the animals from an area that is a fraction of the occupied area. Traps provide a more cost-effective alternative to control chronic infestations, but at low densities attract seastars in from outside the area. Dredges and trawls have been used in Japan to control the seastar prior to seeding an area for shellfish aquaculture, but associated environmental damage would be excessive in an unfished area. Non-specific chemicals, principally quicklime, have been used to locally control seastars on shellfish beds, but collateral damage is high.

Non-specific physical and chemical control may have a role in local control of seastars around aquaculture farms, but sustainable control of the seastar population will require a highly specific biological control agent that can be widely dispersed throughout the population. This possibility is the subject of an on-going research programme.

Edited from a Biocontrol News & Information *News Article, 1999 20 (1), including input from Nic Bax, CSIRO Marine Research, GPO Box 1538, Hobart, Tasmania 7001, Australia; Email: nic.bax@marine.csiro.au*

CASE STUDY 5.20 Mechanical Control Methods for Water Hyacinth

Water hyacinth, *Eichhornia crassipes,* originates in South America, but is now an environmental and social menace throughout the Old World tropics. It impedes water flow and irrigation schemes, disrupts hydro-electric power generation, and impairs water transport and fishing efforts. From a biodiversity perspective, water hyacinth adversely affects water quality, reduces dissolved oxygen levels and increases siltation. It is associated with reductions in fish and aquatic invertebrate populations, and displaces certain aquatic macrophytes. There is growing evidence that water hyacinth changes and significantly speeds up wetland successional processes.

Manual labour to physically remove plants is the oldest means of control. A variety of grabs, knives, hooks and other tools have been developed to do this. However, experience has shown that in many situations where manual removal has been used, the infestation of the weed has grown to such proportions that it is no longer an effective means of control. Nevertheless, appropriately equipped teams from local communities may still be a very effective control method where the weed is not too abundant. However, in many parts of Africa, manual removal of the weed exposes workers to attack by snakes and in some cases crocodiles and there is also a high risk of exposure to water-born diseases such as bilharzia.

The use of booms or barriers to contain or divert the weed as it flows down a river, or to prevent it entering an anchorage or harbour or dam, is widespread. For example, booms have been used to contain the weed in front of the Owen Falls Dam, and at the mouth of the Kagera River where it enters Lake Victoria.

Several countries have used mechanical harvesting of water hyacinth during the last 30 years. Shore-based and floating designs were used in the 1970s and 1980s, having extraction rates of up to 100 tonnes per day (equivalent to about 1.2-1.6 hectares per working day). Newer machines may be able to extract 40 tonnes per hour, but even at these rates of extraction, mechanical harvesting can benefit only those situations in which the amount of weed is relatively limited and it is in a confined and easily accessible area. Several reports show that mechanical control was started when water hyacinth was first found, but eventually proved inadequate to cope with the growth of the water hyacinth.

Studies have shown that costs of mechanical harvesting are on average US$ 600 - 1,200 per hectare, approximately six times more expensive than chemical control using glyphosate. The main advantage to the use of mechanical harvesting is that it removes excessive nutrients and elements from the water body, and may therefore act as a means of slowing or even reversing eutrophication (at least on small lakes). Mechanical harvesting must therefore be linked to a secure system of disposal, either by burning, burial or utilization.

In Egypt, mechanical control of water hyacinth is the sole means of control. Barriers are placed across the river to collect the weed at selected sites, where harvesters mounted on the banks or barges continuously remove it. The efficiency of this operation is undoubtedly greatly enhanced by the fact that it is operating on a large river and irrigation canals, in a situation where the weed collects itself against barrages. Even in this favourable environment, doubts have been expressed as to sustainability of mechanical control alone.

Prepared by Matthew Cock, CABI Bioscience Switzerland Centre, 1 Rue des Grillons, CH-2800 Delémont, Switzerland. www.cabi.org/bioscience/switz.htm.

CASE STUDY 5.21 Chemical Control of *Miconia calvescens* in Hawaii

Hand removal (uprooting) is an effective method of removing plants less than about 3 m tall. Adventitious rooting of uprooted individual occurs occasionally but is rare. If larger miconia cannot be uprooted and are cut down, the stump must be treated with an herbicide (e.g. Roundup, Garlon 4) to avoid resprouting.

An important factor in mechanical and chemical control is the seed bank associated with miconia stands which necessitates monitoring and removal of emergent seedlings for 5-10 years. In large miconia stands, canopy removal often results in a spectacular germination of the miconia seed bank. Miconia seedlings may cover substantial areas in clearings. About 18 months after germination, there can be up to 500-1000 seedlings/m^2, with the tallest about 0.7 m tall. These can effectively be dealt with by spraying with Garlon 3A. A second (and 3rd?) treatment will be required after another 1-2 years to destroy the remaining seed bank (although seedling numbers in subsequent generations are reduced).

The largest miconia population on Maui, discovered from the air in 1993, was initially virtually inaccessible on the ground because of the extremely rough terrain of a 500-year-old lava flow. As a holding action to limit seed production, a helicopter with an attached unit spot-sprayed herbicide on larger, fruiting miconia trees beginning in early 1994. The herbicide (Garlon 4, an ester formulation of triclopyr) was applied with surfactant and blue dye (Turfmark). The dye assisted the pilot in judging application rate and identifying treated plants. Researchers conducted monitoring to assess effects of the spraying. In the initial trials, about 70% of sprayed individuals were killed; others lost leaves and aborted flowers and green fruits, yet recovered and fruited in the next fruiting season. Vegetation plots and tagged individuals are being monitored to determine survival of sprayed miconia, effects on non-target plants, and long-term succession after localized canopy disturbance.

Edited from "Miconia calvescens in Hawaii: a summary" prepared by L. Loope (March 1996), with extensive borrowings from manuscripts by Medeiros, Loope and Conant and by Conant, Medeiros and Loope, and posted on the internet at http://www.hear.org/miconiainhawaii/miconiasummarybylll.htm.

CASE STUDY 5.22 Overview of Successful Rat Eradications on Islands

The most significant cause of extinctions and ecosystem perturbations on islands are introduced species, especially rats (*Rattus* spp.). Complete eradication of introduced rats has been successful on at least 30 islands larger than 10ha. *R. exulans*, *R. norvegicus* and *R. rattus* have all been successfully eradicated. In successful eradications rodenticide bait, usually containing brodifacoum, was spread over all parts of the island either by hand or from a helicopter. Negative impacts on native species were few and short-term. However, none of these eradications took place on islands with extant native rodents and the native biota of most islands had been seriously degraded or exterminated by the rats already.

Eradications were conducted by spreading rodenticide laced bait evenly over the entire island. On many islands bait was placed in bait stations deployed on a grid (usually 50x50 m) and maintained for one to two years. Recently, rats have been successfully eradicated from many islands by aerial broadcasting bait from a helicopter. Additional aerial broadcasting programmes are in progress or planning in New Zealand.

Eradication of rats by trapping alone has been attempted on several very small islands, but was unsuccessful. Subsequent use of rodenticides on these islands was successful.

The rodenticides, brodifacoum, bromadialone and warfarin, have been used alone to successfully eradicate rats from islands larger than 10ha. Brodifacoum has been used most frequently because, unlike warfarin, it can kill rats after a single feeding and resistant rats are extremely rare. Brodifacoum is much more toxic to mammals than to birds, and appears to have almost no effect on reptiles and invertebrates. Therefore, it has been used extensively on islands with no extant native mammals.

Edited from <http://macarthur.ucsc.edu:4000/isla_site/ISLA_SITE.home> "Options for removing introduced black rats (Rattus rattus) *from Anacapa Island, Channel Islands National Park" by Bernie R. Tershy, Donald A. Croll, and Gregg R. Howald.*

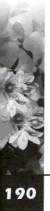

CASE STUDY 5.23 Eradication of the Black Striped Mussel in Northern Territory, Australia

An infestation of the exotic black striped mussel, *Mytilopsis* sp. (also known as *Congeria sallei*) was discovered in Darwin marinas late March 1999. This small, delicate bivalve with a propensity towards fouling (i.e. growing in significant masses on ships' hulls and other places so as to interfere with the flow of water) is a native of tropical and subtropical western Atlantic waters, extending from the Gulf of Mexico to Colombia. *Mytilopsis* sp. has been classified as a serious pest as a result of its potential to cause serious economic and environmental damage. It is believed to have invaded Fiji (prior to 1900), India (ca. 1967) where it has cost the Indian Navy many millions of dollars, and has since been found in Japan, Taiwan (1970s) and Hong Kong (early 1980s).

Despite achieving a maximum size of 2.5 cm in length, an individual mussel is mature at four weeks (ca. 1 cm) and is capable of producing 50,000 offspring. *Mytilopsis* can settle on almost any surface to the exclusion of all other life. At four weeks of age these offspring represent a possible 100 kg of fouling material, which may settle on hulls, chains, ropes, nets, mooring buoys, piles, floating pontoons, inside pipe inlets and outlets, and on any other surface in contact with water. Storm water drains and seawater intakes for industrial plants and mariculture facilities are also vulnerable to fouling by this mussel. In its preferred inshore, low estuarine habitats, introduced populations of black striped mussel are capable of forming mats 10 to 15 cm thick.

Recognising the potential adverse impact on the Australian economy and biodiversity if the bivalve was to become established in Australian waters, the Northern Territory Government (NTG) implemented an immediate containment and eradication programme. Eradication was achieved in the three affected Darwin marinas in 1999 and so far there have been no signs of it reappearing. The operation involved chemically treating the three marinas, surveying and treating 420 exposed vessels (some while at sea), extensive surveys of surrounding waters (with police sharpshooters guarding the divers against crocodiles), 270 people, 2.2 million Australian dollars (excluding labour) and took four weeks.

A monitoring programme has been implemented which documents water quality, and records the presence or absence of marine pests in Darwin marinas and selected sites in the greater Darwin Harbour area. The programme involves the concurrent monitoring of settlement collectors deployed in all four marinas and at selected high traffic sites in Darwin Harbour. The detachable plates and ropes are collected on a regular basis and screened for the presence of aquatic pest species. The plates also provide some indication of the recovery of the marinas.

Complimentary to the settlement collectors is an underwater diver survey, which involves the monthly photography of marked sites within the marinas and harbour. These provide an archival record of the recovery of the marinas. In conjunction with biological samples collected from near the photographed area every three months, the photographs enable more detailed assessment of the recovery of the marinas, and confirmation of the absence of marine pests.

Edited from http://coburg.nt.gov.au/dpif/fisheries/environ/unittext.shtml, supplemented with additional information (e-mail from Nic Bax to Aliens discussion list, 24.5.2000).

CASE STUDY 5.24 Biological Control of an Insect to Save an Endemic Tree on St Helena

In the 1990s, gumwood, *Commidendrum robustum* (Asteraceae), the endemic national tree of St Helena, was in danger of extinction because of an alien insect. Orthezia scale, *Orthezia insignis*, is native to South and Central America, but is now widespread through the tropics. It was accidentally introduced into St Helena in the 1970s or 1980s, and became a conspicuous problem when it started feeding on gumwood in 1991. Gumwood once formed much of the extensive woodland that used to cover the higher regions of the island but is now restricted to two stands of around 2000 trees. It is a typical example of the remarkable indigenous flora on St Helena.

Once the gumwoods became infested in 1991, an increasing number of trees were being killed each year and at least 400 had been lost by 1993. Orthezia scale damages its host primarily through phloem feeding but the colonization of the honeydew that orthezia scale excretes by sooty moulds has a secondary effect through the reduction of photosynthesis. Because orthezia scale is polyphagous, and large populations could be maintained on other hosts such as lantana, it spread easily onto the relatively rare gumwood trees. Gumwoods are susceptible to orthezia scale and if nothing had been done, it is most probable that gumwood would have become extinct in its natural habitat.

The International Institute of Biological Control (now CABI Bioscience) assisted the Government of St Helena to carry out a biological control programme against this pest. There was already an indication that a suitable predator might be available. Between 1908 and 1959, the predatory coccinellid beetle, *Hyperaspis pantherina* had been released for the biological control of *O. insignis* in Hawaii, four African countries and Peru. Substantial control was reported after all the releases.

A collection of *H. pantherina* was obtained from Kenya where it had been introduced to control orthezia scale on jacaranda, and it was cultured and studied in UK quarantine. These studies showed that reproduction of the beetle is dependent on the presence of orthezia scale, that *H. pantherina* normally lays eggs directly onto adult females of *O. insignis* and that the first two instars of the larvae are frequently passed inside the ovisac of the female host, after which the host itself is often consumed. An assessment of the St Helena fauna had also shown that there did not seem to be any related indigenous species (although there were quite a few exotic pest scales present), so that it was concluded that introduction of this predator would not only be safe in terms of effects on non-target organisms, but also would be likely to control the orthezia scale, and save the gumwoods.

In 1993, *H. pantherina* was imported, cultured and released in St. Helena. It rapidly became established and did indeed control orthezia scale on gumwoods. It was concluded that gumwood had been saved from extinction in its natural habitat. This is probably the first case of biological control being implemented against an insect in order to save a plant species from extinction.

Edited from Booth, R.G.; Cross, A.E.; Fowler, S.V.; Shaw, R.H. (1995) The biology and taxonomy of Hyperaspis pantherina *(Coleoptera: Coccinellidae) and the classical biological control of its prey,* Orthezia insignis *(Homoptera: Ortheziidae). Bulletin of Entomological Research **85**, 307-314; and "Saving the Gumwoods in St Helena" by Simon V. Fowler in Aliens (1996) 4, p. 9.*

CASE STUDY 5.25 *Bacillus thuringiensis*, the Most Widely Used Biopesticide

The most widely available and used biopesticides are various formulations of *Bacillus thuringiensis* (known as 'Bt'). Bt is an insecticidal bacterium, marketed worldwide for control of many important plant pests - mainly larvae of the Lepidoptera (caterpillars) but also for control of larvae of mosquitoes and blackflies (Simuliidae). Bt products represent about 1% of the global 'agrochemical' market (fungicides, herbicides and insecticides).

As Bt reproduces, it produces crystalline insecticidal protein toxins. Commercial Bt products are powders containing a mixture of dried spores and toxin crystals, although often the spores are dead and the toxin crystals are the active ingredient. They are applied to leaves or other environments where the insect larvae feed. Bt can only work once the spores have been eaten. The crystal protein is highly insoluble in normal conditions, so it is entirely safe to humans, higher animals and most insects. However, it is solubilised in reducing conditions of high pH (above about pH 9.5) - the conditions commonly found in the mid-gut of Lepidoptera larvae. For this reason, Bt is a highly specific insecticidal agent.

Once the Bt spores and crystalline toxins have been ingested by larvae, the toxin attacks the midgut epithelium causing the formation of holes in cells and thereby creating an influx of ions and water, causing the tissue to break up. When the preparation includes viable spores, these germinate and the bacterium can then invade the host, causing a lethal septicaemia. Death will follow, the speed depending on the amount of Bt and toxins ingested, the size and species of the larvae and variety of Bt used for control. Bt spores do not usually spread to other insects or cause disease outbreaks on their own as occurs with many pathogens.

Initially, Bt was available only for control of Lepidoptera, but screening of a large number of Bt strains revealed some that are active against larvae of Coleoptera (beetles) or Diptera (flies, mosquitoes). The strains active against mosquito larvae have been developed by several companies and used in attempts to control the mosquito vectors of malaria.

The widespread use of these strains has, however, led to the development of resistance in some of the target pests. This may be a major problem as Bt becomes more widely used. The basis of resistance seems to be complex, but one encouraging finding is that, at least in some insects, the receptor for the Bt toxin is an essential gut enzyme, aminopeptidase-N, so any change in this receptor that causes a loss of binding to the toxin could also be detrimental to the insect, potentially reducing the fitness of the resistant insects.

Successful use of these Bt formulations requires application to the correct target species at a susceptible stage of development, in the right concentration, at the correct temperature (warm enough for the insects to be actively feeding), and before the insect pests bore into the crop plant or fruit where they are protected. Young larvae are usually most susceptible. Bt formulations may be deactivated in sunlight and may be effective for only one to three days. Rain can also reduce effectiveness by washing Bt from foliage.

Sources and further information:
http://www.nal.usda.gov/bic/BTTOX/bttoxin.htm,
http://helios.bto.ed.ac.uk/bto/microbes/bt.htm#crest,
http://www.ag.usask.ca/cofa/departments/hort/hortinfo/pests/bt.html,
http://www.nysaes.cornell.edu/ent/biocontrol/pathogens/bacteria.html.

CASE STUDY 5.26   Biological Control of Water Weeds

In the last 50 years, three water weeds of South American origin have stood out as problems in the Old World tropics: water hyacinth (*Eichhornia crassipes*), salvinia fern (*Salvinia molesta*) and water lettuce (*Pistia stratiotes*). All have been the targets for programmes of biological control, each of which has had significant or substantial impact.

These three weeds frequently occur together, and when they do so, water hyacinth normally is the most dominant, and water lettuce is the least dominant. Any of the three species will dominate the indigenous flora and take over calm and slow-moving open water. Accordingly, it is often recommended that biological control of all three weeds should be considered together.

Salvinia molesta was first described from Africa, when it was thought to be a hybrid between the South American *S. auriculata* and an indigenous African species. In 1969-79, initial attempts at biological control by introducing natural enemies from the closely related *S. auriculata* in South America were not very successful. It was only when *S. molesta* was discovered as an indigenous species in south-east Brazil, and the associated weevil, *Cyrtobagous salviniae*, was introduced into Australia in 1980, that successful control was achieved. This weevil has now been introduced into Australia, India, Kenya, Malaysia, Namibia, Papua New Guinea, South Africa, Sri Lanka and Zambia. Everywhere it has been released it has provided effective and often spectacular control of salvinia fern in a matter of months.

Biological control of water hyacinth, native to South America but now an environmental and social menace throughout the Old World tropics, is still the subject of active research. Since 1971, two South American weevils, *Neochetina eichhorniae* and *N. bruchi*, have been widely introduced in Australia, Asia and Africa. In some areas they have provided substantial control, but this is not consistent in different areas. Water nutrient status, average temperature, winter temperatures and other factors probably affect the impact. The search for new insects and pathogens to use as biological control agents continues, and recent discoveries in the Upper Amazon suggest better control may yet be achieved.

The biological control of water lettuce has by comparison proved relatively straightforward. Although there are doubts about the true origin of the plant, its richest associated diversity of natural enemies occurs in South America, and one of these, a weevil, *Neohydronomus affinis*, was selected and introduced into Australia in 1982, giving good control within two years. This success has been repeated in Botswana, Papua New Guinea, South Africa, USA and Zimbabwe.

In the late 1990s there are exciting new reports of the completely successful biological control of another water fern, *Azolla filiculoides* in South Africa, using yet another weevil introduced from America, *Stenopelmus rufinasus*. Clearly there is considerable potential in the biological control of these water weeds from South America, and this should be seen as the option of choice for addressing these invasive weeds in the future.

Prepared by Matthew Cock, CABI Bioscience Switzerland Centre, 1 Rue des Grillons, CH-2800 Delémont, Switzerland. www.cabi.org/bioscience/switz.htm.

CASE STUDY 5.27   Possible Biological Control for European Green Crab

A native of the Atlantic coasts of Europe and northern Africa, the European green crab, *Carcinus maenas*, has invaded numerous coastal communities outside its native range, including South Africa, Australia, and both coasts of North America. It was introduced to the eastern seaboard of the USA some 200 years ago and is frequently held responsible for devastating the soft-shell clam industry in the 1950s in Maine and the Canadian Maritimes. It was first recorded on the West Coast from San Francisco Bay in 1989/90, and since then has moved northwards at an alarming rate of well over 100 miles (160 km) a year. It is considered to be a serious threat to the fisheries and mariculture industry of the Pacific Northwest (with an estimated value of US$45 million/year) and wildlife. Native birds and Dungeness crabs have been singled out as particularly at risk, from predation and/or competition.

In its home range the green crab is found in protected rocky, sandy and tidal habitats. It feeds voraciously, often on bivalve molluscs and particularly mussels, and has a significant impact on populations of these. Preliminary results suggest that it has a similar and perhaps more substantial impact in its introduced range: dramatic declines in other crab and bivalve species have been measured in California and Tasmania, Australia. Native shore crab population declines are greater than 90% in some areas.

A study that compared populations of green crabs from Europe with those from all areas of the world where the green crab has been introduced, found that the introduced populations seemed to be experiencing a release from their natural enemies. Green crabs in introduced populations lacked any parasitism that had direct effects on reproduction, and they reached larger sizes and lost fewer limbs than their European counterparts.

A University of California team is assessing the prospects for introducing a rhizocephalan barnacle, *Sacculina carcini*, that parasitises *C. maenas* in Europe, its native range. This species blocks moulting of its host, and acts as a parasitic castrator, causing female sterility and feminising the males. However, genetic work has shown that putative *S. carcini* from several portunid crab genera in Europe cannot be distinguished genetically, (while they are genetically distinct from other *Sacculina* species). Host specificity will obviously be an important issue when other portunid crabs are present in the proposed release area. Techniques are being developed to assess experimentally the host specificity, and its safety for native crabs. Fortunately, also, the rhizocephalan's life history is such that adding this parasite to a new area could be reversible. Only the female parasitizes and grows in the crab, forming the interna. Unless a second release of parasite larvae is made after females from the first release have ruptured the abdominal wall of the crab and formed the rounded sac, or externa, containing the reproductive organs and brood sac of the parasite, there is no potential for fertilization, and the parasite population would wither away.

Edited from a Biocontrol News & Information *News Article, 1999 20 (1), incorporating input from Armand Kuris, Dept of Ecology, Evolution and Marine Biology, University of California, Santa Barbara, CA 93106, USA. Email: kuris@lifesci.ucsb.edu.*

CASE STUDY 5.28 Control Methods for Australian Pine Include Prescribed Burning

Australian pine, *Casuarina equisetifolia*, is native to Malaysia, southern Asia, Oceania and Australia. It is a deciduous tree with a soft, wispy, pine-like appearance that can grow to 100 feet or more in height.

Australian pine was introduced to Florida in the late 1800s and planted widely for the purposes of ditch and canal stabilization, shade and lumber. It is now established in the Hawaiian and other north-east Pacific islands, coastal Florida, Puerto Rico, the Bahamas, and many Caribbean islands. It is fast-growing, and produces dense shade and a thick blanket of fallen leaves and hard, pointed fruits, that completely covers the ground beneath it. Dense thickets of Australian pine displace native dune and beach vegetation, including mangroves and many other resident, beach-adapted species. Once established, it radically alters the light, temperature, and soil chemistry regimes of beach habitats, as it outcompetes and displaces native plant species and destroys habitat for native insects and other wildlife. The ground below Australian pine trees becomes ecologically sterile and lacking in food value for native wildlife.

For new or small infestations, manual removal of Australian pine seedlings and saplings is recommended. For heavier infestations, application of a systemic type herbicide to bark, cut stumps, or foliage is likely to be the most effective management tool. Prescribed fire has also been used for large infestations in fire-tolerant communities.

Edited from http://www.nps.gov/plants/alien/fact/caeq1.htm, Casuarina equisetifolia *L. by Jil M. Swearingen, U.S. National Park Service, Washington, DC. For more information, see* http://tncweeds.ucdavis.edu/esadocs/casuequi.html.

CASE STUDY 5.29 An IPM Research Programme on Horse Chestnut Leafminer in Europe

Horse chestnut, *Aesculus hippocastanum*, is widely planted as an ornamental and amenity tree in Europe. In the 1980s, a new leafminer pest of unknown origin appeared in the Balkans, and since then has spread from there into Central Europe. The leafminer, a small moth, was described as *Cameraria ohridella*, new to science, but assumed to be an alien introduction from an unknown location. The infestation became so severe that, in addition to harmful ecological impacts, economic damage was also reported, since this tree species is the prevailing shade tree in restaurant and bar gardens in Central Europe. A group of European scientists put together a successful European Union project proposal "Sustainable control of the horse chestnut leaf miner, *Cameraria ohridella* (Lepidoptera, Gracillariidae), a new invasive pest of *Aesculus hippocastanum* in Europe" to study control options for this new invasive alien species in Europe. The key elements of the programme are:

➤ To assess the physiological and economical impacts of the moth.

➤ To monitor the presence and impact of *C. ohridella* in natural stands of *A. hippocastanum* in the Balkans.

➤ To assess the potential host range of the moth.

➤ To study the chemical interactions between the moth and its host plant.

➤ To develop pheromone-based monitoring and control methods.

➤ To determine the area of origin of the moth, define the factors controlling the moth in its natural environment and evaluate the potential of exotic natural enemies as biological control agents in Europe.

➤ To evaluate the natural enemy complexes in Europe at continental level, assess the potential of European natural enemies to naturally control the pest and develop control techniques involving the conservation of these natural enemies.

➤ To assess the efficiency of presently used cultural control methods, study the possibility of improving their efficiency, and of modifying them to conserve or augment the action of natural enemies.

➤ To develop mapping methods to estimate damage and dieback risks.

➤ To study the epidemiology of the moth, dispersal mechanisms and dispersal prevention methods at the western fringe of its distribution.

➤ To integrate pest risk assessments, monitoring and control methods into IPM strategies adaptable to various geographic, economic and climatic regions.

➤ To use the invasion of *C. ohridella* as a case study for recommendations on control of other invasive insect tree pests in Europe.

Prepared by Marc Kenis, CABI Bioscience Centre Switzerland, 1 Rue des Grillons, CH-2800 Delémont, Switzerland, http://www.cabi.org/BIOSCIENCE/switz.htm.

CASE STUDY 5.30   Integrated Management of Water Hyacinth

The problems caused by water hyacinth (*Eichhornia crassipes*) are multi-faceted (see Case Study 5.1 "Problems Caused by Water Hyacinth as an Invasive Alien Species"). As a result the objectives of many control programmes may be unclearly defined. Effective management plans are needed which involve all local stakeholders in their development, together with inputs from specialists in all aspects of weed control and utilization. The principal options for control of water hyacinth are mechanical, chemical and biological control. Utilization should not be considered an effective control strategy by itself but is an important consideration for an integrated control programme.

Biological control is the only permanent and sustainable control option, and as such it must be the basis of any control programme. It has proved to be an adequate control method on its own in several instances in developing countries (e.g. Sudan, Papua New Guinea, Benin). Using currently available agents, it usually reduces biomass by 70-90%. The principal drawback with biological control of water hyacinth is the time required to achieve control. In tropical environments this is usually 2-4 years and is influenced by the extent of the infestation, climate, water quality, and other control options. Because of the time taken to achieve full impact, biological control should be pursued as a matter of the greatest priority as soon as an infestation of the weed appears. Other control options will need to be integrated with the biological control.

As the weed infestation increases, the capacity of the biological control agents to control it effectively and quickly diminishes, so that other interim means of control may be needed. Herbicides have been used extensively around the world as a quick and effective means of controlling water hyacinth. They are relatively cheap, with costs per hectare for aerial application of US$25-200. Studies of herbicide residue levels and the environmental impact of the weed upon aquatic communities and fish suggest that if used correctly, both glyphosate and 2,4-D may be safely used in tropical wetland communities. There may be some prospects for using barriers and tows to position mats in optimal places for spraying. Major drawbacks of using herbicides are that they are non-selective and could cause major environmental problems if incorrectly applied. Also, chemical control needs to be carried out repeatedly as the weed quickly re-grows in tropical environments.

Chemical control can be effectively integrated with biological control by spraying only a part of the water hyacinth infestation. The timing of the spray should be judged so as to coincide with peak numbers of the adult dispersive stage of the biological control agents. These may then colonize the unsprayed plants thereby maintaining the biological control.

Mechanical removal of the weed is used in several countries (see Case Study 5.20 "Mechanical Control Methods for Water Hyacinth"). Mechanical removal with harvesters is about six times more expensive than chemical treatment. It is also slow and hence, not suitable for clearing large mats of the weed. However, it is the most effective means of control in critical areas, such as hydro-electric dams and ports, where confined areas become choked with the weed. After removal to the shore, the weed must be disposed of effectively and safely to prevent plants and seeds returning to the water.

Prepared by Matthew Cock, CABI Bioscience Switzerland Centre, 1 Rue des Grillons, CH-2800 Delémont, Switzerland. www.cabi.org/bioscience/switz.htm.

CASE STUDY 5.31   What Can Happen When an Invasive Alien Species is Controlled

Waikoropupu Springs in New Zealand is a remarkable freshwater spring ecosystem of about one hectare, of great importance biologically and culturally. For example, in these springs there is a moss species (*Hypnobartlettia fontana*) that grows nowhere else.   The springs were heavily invaded by watercress (*Rorippa nastustrium-aquaticum*), an introduced species, which grows to great bulk in water up to six metres deep (the maximum depth of the springs) and totally smothered the springs and most of its unique communities.

Around 1990 the need to control the watercress was recognized. Previously cattle had access to the area and may have controlled the watercress but the area is now fenced. Watercress seldom grows to this size and is rarely considered troublesome to this extent elsewhere in New Zealand.

Control methods in the springs were considered and a hand-weeding programme was introduced as the most practical and environmentally acceptable solution. In some of the areas left bare by watercress removal the native aquatic species did recover. In other areas native bryophytes and algae established and then an introduced rush, *Juncus microcephalus*, invaded the bare areas and recovering communities, and now appears to be invading the limited remaining areas of native vegetation as well. This species is much worse than watercress as it has a stronger root system and its removal causes much more disturbance. Worse still, two introduced aquatic grass species, *Glyceria fluitans* and *G. declinata* have recently (2000) been identified in the springs. They appear to have stronger root systems again.

While the Department of Conservation intends to control the exotic rush in the springs to some extent, plans are yet to be agreed, and whatever they do, will be done very carefully, with detailed monitoring to try and prevent worse invasions.

Edited and adapted from an e-mail sent to the Aliens list-server, 31 July 2000, by Melanie Newfield, Weed Ecologist, Department of Conservation, Nelson/ Marlborough Conservancy, Private Bag 5, Nelson, New Zealand.

CASE STUDY 5.32   Development of a European Research Programme on Horse Chestnut Leafminer

Horse chestnut, *Aesculus hippocastanum*, is widely planted as an ornamental and amenity tree in Europe. In the 1980s, a new leafminer pest of unknown origin appeared in the Balkans and has spread from there into central Europe. The leafminer, a small moth, was described as *Cameraria ohridella*, new to science, but assumed to be an alien introduction.

The larvae of this moth tunnel within the leaves of its host, causing unsightly markings, leaf loss, and overall ill-health of the horse chestnut trees. Since the first observations of the moth in Macedonia in 1984, Austria in 1989 and Bavaria in 1992, this pest has attracted more public attention in these regions than any other tree pest in the history of forest entomology. As a result, several scientific teams started to work on the moth. However, until very recently this research was done without any co-ordination at the European level, some investigations being duplicated whereas other aspects were totally neglected.

The leader of one of the teams working on *C. ohridella* at the University of Munich, Germany, built a team to apply for funding to the EU 5th framework. He contacted:

➤ The co-ordinator of *C. ohridella* research programmes in Austria, the country where most work had been done so far;

➤ The Institute of Organic Chemistry and Biochemistry, Czech Republic, where scientists had just discovered the pheromone of *C. ohridella*;

➤ The CABI Bioscience Centre Switzerland, where a scientist had published on the possibilities for classical biological control of tree pests such as *C. ohridella* in Europe and had been collaborating with the University of Munich through field collections and publications;

➤ A scientist from the University of Bern with experience in parasitoid and predator ecology who had just started to study *C. ohridella* in Switzerland; and

➤ A scientist from the University of Trieste, who was already in contact with the University of Munich and was interested to work on a rather different aspect - the effect of the pest's damage on the trees' water balance.

The group first met in Germany in 1999 to establish the proposed work programme and decided to include other teams, to better balance the project, both geographically and scientifically:

➤ INRA Orléans, France, in order to have a French team studying the dispersal and epidemiology of the moth as it enters a new area, i.e. France; and

➤ The University of Drama, Greece, and the University of Sofia, Bulgaria, to incorporate partners in Balkan countries, since this is the only area in Europe with indigenous horse chestnut forests, which the participants were planning to study.

The writing up of the proposal was a collaborative effort between the partners, each contributing sections reporting on their own work plan, and the co-ordinator compiling the information. One of the problems encountered was that none of the team members were native English speakers, but the team members more fluent in English were able to address this. It is worth pointing out that the deadline for submission would not have been met without E-mail correspondence and the establishment of an FTP (file transfer protocol) server at the University of Munich that facilitated the exchange of files among participants. The project was accepted for funding under the EU 5th Framework in 2000.

Prepared by Marc Kenis, CABI Bioscience Centre Switzerland, 1 Rue des Grillons, CH-2800 Delémont, Switzerland, http://www.cabi.org/BIOSCIENCE/switz.htm.

CASE STUDY 5.33 Social and Environmental Benefits of the Fynbos Working for Water Programme

The "Fynbos Working for Water Program" is a sub-programme of the South African Department of Water Affairs and Forestry's "Working for Water" Program. This name refers to the jobs being created to clear water catchment areas and river courses of woody invasive alien plants. The programme is of enormous benefit to the environment but also has clear socio-economic benefits. A major problem in the young democracy of South Africa is unemployment and related social problems such as crime. The social aims of the programme are the empowerment and upliftment of rural communities.

South Africa, and especially the Western Cape Province with its unique fynbos vegetation (which forms part of the Cape Floral Kingdom), has an enormous problem with invasive alien trees and shrubs. Fynbos is a fire-prone vegetation type that is highly susceptible to invasion by alien plants. Several species from the Mediterranean Basin, North America, and especially Australia, are major problems. Species such as *Pinus pinaster* (Mediterranean Basin), *Pinus radiata* (California) and *Hakea sericea* (Australia) are a major threat to the fynbos in the mountainous areas of the Western Cape Province, whereas Australian *Acacia* species such as *A. mearnsii* and *A. saligna*, and *Eucalyptus* spp. are threatening the lowlands and riparian areas. Because of extensive budget cuts during the political transition of South Africa, the invasive alien plant clearing programme had come to a virtual halt.

An informal discussion group, the Fynbos Forum, comprising scientists and environmental managers, held a workshop in November 1993 to discuss the effect of invasive alien plants on runoff from fynbos catchments. They adopted a resolution to develop a "road show" presentation, to demonstrate to policy makers the effect of invasive alien plants on both water runoff and biodiversity, and the potential socio-economic consequences of this. The "road show" was presented to Kader Asmal, Minister of Water Affairs and Forestry, in July 1995. The Minister saw the potential of the project as a tool in the Reconstruction and Development Program of South Africa.

In September 1995, R25 million (US$5.5 million) were allocated to the national programme, with R13.5 million of this going to the 1.14 million hectares of fynbos catchments of the Western Cape Province. Invasive alien plants occur in almost half of this area. Of the total invaded area, more than 60,000 ha are covered with alien plant stands having canopy cover of 25–100%. Between the start of the Working for Water Program in October 1995 and the end of August 1996, 39,000 ha had been cleared, including nearly 7,000 ha of dense stands (having > 25% canopy cover). The Fynbos Working for Water Program employed more than 3000 people at its (first) peak in March 1996. More people are now being employed following the injection of a further R40+ million into the project. Alien plant control is not a once-off job. For the Fynbos Working for Water Program to be successful it will have to follow up the initial clearing operations at regular intervals for 8-10 years to ensure that the seed banks are depleted.

In this programme, short-term social benefits contribute towards the realization of long-term development and environmental goals.

Edited from "The Fynbos "Working for Water" Programme" in Aliens *(1997) 5, p. 9-10, by Christo Marais, Programme Manager, and Dave Richardson, University of Cape Town.*

CASE STUDY 5.34   Ecotourism as a Source of Funding to Control Invasive Species

The particular threats facing biodiversity-rich small island states by invasive alien species have been widely recognized. Taking this into consideration it is all the more important for such island states to recognize their strengths and the primary economic forces at work in their countries and focus these to the benefit of the conservation and sustainable use of biodiversity, for example through management of invasive species.

In Seychelles, rats (*Rattus rattus* and *R. norvegicus*) have probably had more impact upon the endemic biodiversity than any other factor. For example within the central Seychelles alone (41 islands) there are six species and one sub-species of endangered land birds. Of these, only one is believed to be able to co-exist with *R. rattus*. It is consequently a national priority to mitigate the impacts of rats. Cats have also been very damaging, and where cats and rats co-exist, they both need to be controlled. Rat and cat eradication is expensive, requiring the use of helicopters to deploy bait, and specialized expertise from overseas to implement the project. Consequently the cost per island is high and in a new initiative to eradicate these species on key offshore islands, it was essential to incorporate more than one island into the programme.

The primary economic force in Seychelles, as with many small island developing states, is tourism. The challenge faced, therefore, was how to capture and utilize the resources of tourism in Seychelles to meet the goals of rat-impact mitigation.

In this case the unifying factor is eco-tourism. At present, there are already two islands in Seychelles, both reserves, that are financed exclusively by eco-tourism. The concept and viability of such operations is recognized in-country.

A shortlist was prepared of islands which would have potential for species reintroduction if their rats and cats were eradicated, i.e. of sufficient size and potential habitats for species introduction, controlled access to island, sufficient distance from neighbouring islands etc. The Government negotiated with the islands' owners or management bodies, and proposed the endowment of legal protected status to islands that would establish and maintain predator-free status. Rat prevention protocols have been developed to meet the specific circumstances of each island, but in general involve rat-proof containers for shipment of supplies, a rat-proof room for opening all imported packages, no landing of craft other than simple open-deck dinghies with no compartments and regulations regarding mooring distance for sea-going craft, together with ongoing monitoring through placement of "chew-sticks" and line trapping to enable early detection of any new introductions.

With the lure of potential future eco-tourism revenue, three islands agreed to be incorporated in the programme. Two of these islands are private with exclusive hotel operations, the third is a National Park managed by the Marine Park Authority.

The logistically very complex programme of eradications was undertaken May – August 2000 with the private islands funding their own operations, with overall costs approaching US$250,000.

Prepared by John Nevill, Director of Conservation, Ministry of Environment and Transport, Republic of Seychelles. E-mail: chm@seychelles.net

The following example has been edited from an e-mail advertisement that was recently circulated:

WILDLIFE VOLUNTEERS NEEDED FOR WEED MANAGEMENT PROJECT IN MAURITIUS WITH THE MAURITIAN WILDLIFE FOUNDATION

The Mauritian Wildlife Foundation is a highly successful Non-Governmental Organization working to save the globally endangered flora and fauna of Mauritius and Rodrigues. Notable successes include the conservation of the Mauritius kestrel, pink pigeon and echo parakeet and ecosystem restoration projects in Mauritius, its offshore islets and Rodrigues.

Plant Conservation Projects
We are seeking a motivated plant ecologist with an interest in invasive species management to manage field trials & set up new surveys to aid our weed management programme on Ile aux Aigrettes.

Background to the Ile aux Aigrettes restoration project
The greatest current threat to the native biota of Mauritius is the action of invasive exotic plants and animals. Many of the strategies of in-situ conservation of native species in Mauritius focus on the control of exotics, etc. …….
Weed control field trials are ongoing. We have set up field trials in intensively managed areas where we are comparing manual, chemical and integrated control strategies. We require a volunteer to continue the monitoring of these trials. We also would like the person to set up further trials on individual species. The project will NOT involve long hours of manual weeding!!!! We have long term staff available for this.

Requirements
Applicants should be able to key plants to genus and species and have some experience of plant propagation methods and ecological methodologies. They must be willing to learn, work in a team, maintain a good attitude, have the ability to walk across rough terrain and work long field hours. You will also need a driving license. Minimum age 21.

Experience gained
You will gain practical experience of weed management, nursery management, ecosystem restoration, working as a team member, and group dynamics.

Expenses & living conditions
We cannot pay travel costs to and from Mauritius. Accommodation and relevant field equipment are provided. You will need binoculars and field clothes. We have a field station on Ile aux Aigrettes and a house on the Mauritian mainland for aviary staff and for rest & recuperation for other field staff). We generally advise people to bring the equivalent of about 200 pounds sterling monthly living expenses.
Volunteers often go on to pursue post-graduate degrees following on from the project they were working in or in a linked area. There can be no guarantee of this.

What to expect
Long field days, tropical sun, tropical seas and tropical rain, meeting new people, and learning about new places, some of the rarest birds, plants, reptiles and bats in the world.
Position open until filled. APPLY ASAP.
Send a covering letter, a CV and letters of recommendation (or names of referees to contact). Please include the dates you are available.

CASE STUDY 5.36   Using the Media to Create Awareness and Support for Management of Invasive Species: the Seychelles Experience

It is tempting for small island states to focus upon the limitations and difficulties they face, particularly in terms of infrastructure and logistics and the strictures these place upon co-ordinated and sustained operations.   However, one must always strive to utilize national characteristics to advantage.

The Republic of Seychelles has some 115 islands spread over an Exclusive Economic Zone of 1.3 million square kilometres and a population of approximately 80,000.   As such Seychelles faces all the difficulties associated with the stereotyped small island developing state scenario.

The population of Seychelles, is relatively affluent and 92% of households have a television (Ministry of Information and Culture, Unpublished Report 2000).   In Seychelles, there is only one television station, a fact often bemoaned by the populace for the lack of choice they therefore experience.

Of course, the limitation of one channel has the advantage of a captive audience and unparalleled access to the population particularly during peak viewing hours.   Hence television is a very powerful tool in raising public awareness.   Despite this lack of choice, programmes still need to be to a satisfactory standard and well presented, with the relevance to the average viewer highlighted in order to maintain public interest.

In Seychelles, children are targeted separately with a special weekly programme called "Tele-zenn" which loosely translated means "Youth TV".   This programme is presented by children, for children and addresses environment issues from their perspective and in their native language. With specific regard to invasive species successful awareness campaigns have been carried out regarding invasive creepers, pond plants, the introduced barn owl and release of potentially invasive cage birds, notably parrots.   These campaigns have resulted in considerably increased co-operation from the public with regard to introduction and/or control of these pest species.

Prepared by John Nevill, Director of Conservation, Ministry of Environment and Transport, Republic of Seychelles. E-mail: chm@seychelles.net

CASE STUDY 5.37 Community Participation in Control of Salvinia in Papua New Guinea

Salvinia molesta, is an aquatic floating fern from South America, capable of forming dense mats by vegetative growth as an introduced species in the Old World and USA. It was the target of a series of very successful biological control programmes in the 1980s (see Case Study 5.26 "Biological Control of Water Weeds").

In Papua New Guinea, the impact of salvinia was particularly severe in the Sepik River, which drains much of the north-eastern part of the island of New Guinea. The lives of the people of the region are linked very closely with the river, which provides their main source of food and the principal means of travel in an area lacking roads. Complete domination of much of the open water, particularly fishing grounds in oxbow lakes and the margins of all water bodies must have displaced much of the indigenous flora and fauna, although this was not systematically documented. The impact on the lives of the indigenous people was very clear, and highlighted by cases of people who were unable to reach medical assistance because of the infestations. Some villages were abandoned when they became inaccessible by canoe.

When a biological control programme was implemented in 1982-85 by United Nations Development Programme with the assistance of CSIRO Australia, establishment of the biological control agent, *Cyrtobagous salviniae*, was rapidly achieved in lagoons close to the project base at Angoram on the lower Sepik River. The challenge then was how to redistribute the weevils to the rest of the river system. Redistribution was easy in principle since bags of salvinia fern together with weevils could be collected from the infested lagoons and simply released into other affected parts of the watershed. In practice the lack of infrastructure made this very challenging.

Messages were sent out via radio suggesting that villagers further up river could visit the infested lagoons and collect bags of material (salvinia with weevils) and take them back to their water bodies and release them. This was done and canoes were used to ferry infested salvinia up the river. A single engine aircraft was also used to ferry infested salvinia longer distances from Angoram to mission airstrips near the river or lagoons. Mission staff and local people then organized its transfer and distribution to affected water bodies.

The involvement of the main stakeholders in the Sepik River in this way ensured that the biological control agents were distributed much more quickly than relying on a central distribution system. The resultant rapid control of the alien weed is one of the most successful stories of biological weed control.

Prepared from inputs by Peter Room and Mic Julien, CSIRO, Brisbane.

CASE STUDY 5.38 The Use of Local Part-time Volunteers to Help Restore a Nature Reserve on Rodrigues

Rodrigues is a tiny island 550 km to the east of Mauritius, in the Indian Ocean. Politically it is part of Mauritius. Rodrigues currently has the dubious distinction of being one of the most degraded tropical islands in the world. All mature forests are dominated by invasive alien species and no contiguous areas of full canopy native forest remain. From 1996 the Mauritian Wildlife Foundation (MWF), a conservation NGO with the goal of saving the endangered biodiversity of Mauritius and Rodrigues, together with the Rodrigues Forestry Service began the work of restoring the 10 ha nature reserve of Grande Montagne to full canopy native forest. This work comprises of the gradual clearance of dense stands of alien trees, the replanting of the cleared areas with a diverse array of nursery-grown natives and the maintenance of these plantations.

Initially all the work in the reserve was carried out by Forestry and MWF staff (including full time Rodriguan and expatriate volunteers). Since mid-1999 involvement in the restoration has been opened up to part-time volunteers. Up to 30 people come to work in the reserve every Saturday and in the school holidays. They are usually from pre-existing groups such as the scouts, secondary schools and professional organizations and are organized by their respective group leader. Initially a group helps with the overall work in the reserve. If they continue to show keenness the group is encouraged to adopt a plot in the reserve, which then becomes 'their plot'. This plot is cleared by the group, who undertake to manage it as long as is necessary. One group has chosen to monitor the success of their work by conducting vegetation surveys in permanent quadrats within their adopted area.

A full time member of the MWF team supervises all the work so that there is full technical backup and the work is consistent with the overall management aims for the reserve. MWF provides all the plants from its nursery so that all material used is of a known provenance. The team finishes off most days with a session on some aspect of the conservation work such as the identification of particular native and non-native plant species, the reasons for the conservation methods used and aspects of other conservation programs in Mauritius and Rodrigues.

The work with local part-time volunteers is continuing to attract new support. Some of the key reasons for the success of the programme are:

➤ A high level of awareness of the Rodrigues biodiversity restoration project raised by the work of the Rodrigues Community Educator.

➤ Working through pre-existing community groups who organize their members themselves.

➤ Easy access to the reserve from a nearby public road.

➤ The ownership the groups feel of their part of their nature reserve.

➤ The opportunity to learn about the natural heritage of Rodrigues.

➤ Reliable leadership in the field provided by dedicated conservationists.

➤ The support of the local administration for the project.

Prepared by John Mauremootoo, Mauritian Wildlife Foundation, 4th Floor Ken Lee Building, Edith Cavell Street, Port Louis, cjmaure@intnet.mu

CASE STUDY 5.39   A Preliminary Risk Assessment of Cane Toads in Kakadu National Park

The cane toad, *Bufo marinus*, introduced to Australia in 1935, will soon arrive in Kakadu National Park (KNP), a World Heritage area with Ramsar-listed wetlands. Cane toads eat a wide variety of prey, have greater fecundity and develop more quickly than native frogs and toads (anurans), and possess highly toxic chemical defences against predators. They tolerate a broad range of environmental and climatic conditions, and can occupy many different habitats. To date, no effective control methods for cane toads have been developed. There is concern that the status of KNP could be diminished if cane toads negatively affect any of the Park's natural and cultural values. Consequently, an ecological risk assessment was undertaken to predict key habitats and species at risk, from which recommendations for new monitoring programmes could be made, the relevance of existing programmes evaluated, and some management options identified.

The approach, based on a wetland risk assessment framework developed for the Ramsar Convention on Wetlands, involved identification of: the problem; the (potential) effects; the (potential) extent of the problem; the subsequent risks; and the information gaps. The outcomes were used to provide advice for monitoring and risk management.

A total of 154 predator species were listed. Ten species were in risk category one (i.e. the greatest risk category), with northern quoll (*Dasyurus hallucatus*, a carnivorous marsupial) being assigned highest priority. The nine remaining species were assigned high priority. Twelve species or species groups were in the second risk category, while the third risk category contained 132 species or species groups. Risks to prey species were difficult to predict, but those most likely to be affected included termites, beetles and ants. Similarly, risks to potential competitor species were unclear, but potential effects on some native frog species and insectivorous lizards were of concern. A great deal of uncertainty surrounded the prediction of risks to the environment. Contributing to this was a lack of understanding or quantitative data on i) impacts of cane toads on animal populations; ii) populations, distributions and general ecological information on the native fauna of KNP; and iii) cane toad densities within the Kakadu National Park.

Seven priority habitat types were identified for monitoring: floodplain communities; swamp communities; monsoon forest; riparian communities; woodland and open forest communities; springs, soaks and waterholes; and escarpment pools. Priority species for monitoring included northern quoll, the varanid lizards, several elapid snakes and dingo. Other species warranting close attention included some small mammals, ghost bat, black-necked stork, freshwater crocodile, and a range of native frogs. With a few exceptions, it was concluded that historical or current monitoring programmes within KNP were unsuitable for providing a baseline for the assessment of toad impacts. Finally, monitoring and research recommendations to address critical information gaps were also made.

Cane toad control options are extremely limited. It was suggested that particular, sustained measures may prove effective in localized areas (e.g. townships, caravan parks), but that broad scale control is not possible, as chemical and biological control methods are insufficiently developed at this stage. It was recommended that Parks North manage the invasion of cane toads initially by i) ensuring that monitoring efforts are underway to assess impacts of cane toads to the Kakadu National Park, and ii) investigating measures by which cane toads can be managed on a localized basis.

Edited from: van Dam, R.A.; Walden D.; Begg G. (2000) A preliminary risk assessment of cane toads in Kakadu National Park. Final Report to Parks North. Supervising Scientist, Darwin, N.T., 89 pp.

CASE STUDY 5.40   Community-based Aboriginal Weed Management in the 'Top End' of Northern Australia

In the northern part of the Northern Territory of Australia, known as the 'Top End', Aboriginal people own a large area of land (over 170,000 km^2) including approximately 87% of the Northern Territory coastline. They rely heavily on these lands for food, for cultural reasons and, increasingly, for economic independence. Apart from its cultural significance the land is also host to a large portion of Australia's biodiversity. There are several threats to the integrity of this land such as changing fire regimes and invasion by feral animals and weeds, in particular the rampant Central American floodplain weed *Mimosa pigra* (mimosa).

Unfortunately Aboriginal groups have a low capacity to deal with such new and emerging threats to their land. Traditional ecological knowledge and land management skills do not adequately address such problems and weed control, in particular, has often been given a low priority because the potential environmental impact of particular weeds is not fully recognized. Aboriginal people have limited personal resources and the resources of their representative organizations have been focussed on other priority issues, such as claiming back land and the provision of housing, water, electricity etc.

The Caring for Country Unit (CFCU) of the primary representative organization for Aboriginal people in the Top End, the Northern Land Council, is using the weed mimosa to 'kick start' formalized weed and land management in a number of key areas on Aboriginal lands across the Top End. The project involves contributions from a range of agencies, and aims at developing a spirit of multi-agency collaboration to strategically address weed management and other land management and community issues. The project could result in major conservation benefits, increased employment in communities and the eventual development of enterprises based on natural resources.

The project focus is on assisting communities to build their capacity to undertake land management work for themselves. Participants are employed on the Community Development Employment Program, a Commonwealth Government employment program for Aboriginal people and basic training and resources are brokered to initiate mimosa control work. Emphasis is placed on prevention and early intervention. Over time, with increased experience and confidence and through more broad-based training the work is broadened to include other land management issues.

Attendance at workshops and participation in field trips help people better understand the concepts of integrated conservation and development. Communities are now investigating enterprise development based on the sustainable use of natural resources that could, in time, help fund land management activities.

CFCU does not seek to develop generic models for land management, recognizing that community needs, capacity, aspirations and community structures will vary across the region. The over-arching goal is to assist Aboriginal landowners and communities to develop locally appropriate formal land management programmes where informal traditional land management is inadequate to address emerging problems. No single model for a formal land management programme has been specified and nor is it intended to develop such. Empowerment is the key.

Prepared by Michael Storrs, Wetlands Officer, Northern Land Council, PO Box 42921, Casuarina, NT, 0811, Australia. Email: *michael.storrs@nlc.org.au*

CASE STUDY 5.41   Invasive Species Mitigation to Save the Seychelles Black Parrot

The Seychelles Black Parrot (*Coracopsis nigra*) is critically endangered, and restricted to the islands of Praslin and La Digue within the Seychelles Archipelago. Its initial decline following human colonization was believed to be due to hunting (it was considered a pest of fruit trees) and loss of breeding sites (dead palm tree trunks) due to forestry management. An intensive study was started in 1982, which determined that breeding success in the remaining population was extremely low. This low success was due to rat predation of begging chicks in the nest. Extensive rat control measures i.e. trapping and tree protection did not greatly alleviate the impact due to the abundance and arboreal nature of *Rattus rattus*.

This, combined with the lack of suitable nesting sites, made the status, of what was likely an aging population, critical. A total population crash seemed to be likely which would lead to extinction of the species. Consequently an original nest-box design (Mr. Victorin Laboudallon, Conservation Officer) was developed to exclude rats.

The design is complex and expensive utilising a concrete foundation, a two metre length of galvanized metal pipe topped by a perpendicular metal plate, upon which the nest-box sits. The construction has to be sturdy because the entrance to the nest-box consists of a length of rotten hollow palm trunk some two metres in length placed on top to provide the natural and favoured appearance of the nesting site. Furthermore, nest-boxes have to be carefully sited to avoid rats jumping on to the box from overhanging tree branches, and man-made firebreaks have proven most suitable for this purpose.

Ten boxes were set up for a trial period of several years to test the efficacy of the design. One box in three, on average, was occupied and breeding was very successful. Following on from this promising trial, a project to build and install a further one hundred nest-boxes has been started.

Prepared by John Nevill, Director of Conservation, Ministry of Environment and Transport, Republic of Seychelles. E-mail: chm@seychelles.net

CASE STUDY 5.42   Eradication of the Grey Squirrel in Italy: Failure of the Programme and Future Scenarios

The American grey squirrel (*Sciurus carolinensis*), introduced into the British Isles and Italy as a pet species, when naturalised causes severe damage to forests and commercial tree plantations by bark-stripping, replaces the native red squirrel (*S. vulgaris*) through competitive exclusion, and is also suspected of being a source of parapoxvirus, lethal to the red squirrel. Italy has the only populations of grey squirrels living in continental Europe and their expansion is expected to cause an ecological catastrophe at a continental scale as already experienced in the UK. The grey squirrel was introduced into Piedmont (North-West Italy) in 1948 and rapidly became established. For several decades, the grey squirrel was recorded only close to the release site, but from 1970 it started to spread into the surrounding area.

From 1989, several international organizations and scientists, including the IUCN and the British Forestry Commission, advised the Italian authorities of the threat that the grey squirrel posed to the red squirrel, and urged its eradication. The National Wildlife Institute (NWI) approved a recommendation to eradicate the grey squirrel from Italy, and warned the Ministry of Environment, the Ministry of Agriculture, and all local administrations (responsible for pest management plans) about the drastic expansion of the grey squirrel's range and the risks related to its presence.

By 1996, the grey squirrel had greatly expanded its range and was predicted to reach the Alps in about two years. By means of drey counts and capture-recapture censuses, the total population size was estimated at 2,500-6,400 individuals at that time. In view of the urgency of removing the grey squirrel from Italy, in 1997 the NWI in co-operation with the University of Turin produced an action plan for eradication. One of the first steps of the plan was the experimental removal of the small population present in the Racconigi Park, in order to test effective and humane techniques. The local authorities would carry out further eradication efforts. The project plan was sent to all the main Italian NGOs and, on the basis of the resulting comments, the following protocol was adopted: 1) live-trapping of the squirrels, in order to avoid risks for non-target species; 2) frequent checking of traps, to reduce the captivity period; 3) anaesthesia with halothane, a tranquilliser that reduces stress in rodents; 4) subsequent euthanasia of animals with an overdose of halothane; and 5) constant supervision by a veterinarian. On the basis of the revised protocol, most NGOs approved the eradication plan, and the trial eradication started in May 1997. The preliminary results were very encouraging. During just eight days of trapping at Racconigi, 188 animals (> 50% of the estimated population) were trapped and euthanased. The adopted procedure of euthanasia resulted in a significant reduction of stress to the squirrels: they reached unconsciousness in less than a minute and could be euthanased in the field, with very limited manipulation.

However, some radical animal rights groups strongly opposed the project, organising small demonstrations at a local level. Then, in June 1997, they took the NWI to court and managed to halt the project. The case was closed only in July 2000, with the full acquittal of the NWI. The three-year legal struggle caused the failure of the entire campaign. The enforced early termination of the trial eradication did not allow completion of the pilot programme and local administrations did not proceed with the planned eradication. As a result of the suspension of all action, the grey squirrel has now reached the forests of the Alps and eradication is no longer considered feasible. Expansion into a large part of Eurasia, and subsequent decline of the red squirrel, is the likeliest scenario.

Prepared by Piero Genovesi, National Wildlife Institute, Via Ca' Fornacetta 9 - 40064 Ozzano Emilia (BO), Italy, email: infspapk@iperbole.bologna.it

CASE STUDY 5.43   Students Help to Restore a Rainforest by Weeding

Singapore has an ongoing programme to clear alien species, especially exotic plant species in its rainforest. Over the years, exotic creepers have been smothering the rainforest edges and are an insidious threat to the remaining rainforest in Singapore. The creepers are strangling and causing the death of old and mature trees. They also kill off young native saplings by smothering them and stunting their growth, preventing regeneration.

The National Parks Board (NParks) has carried out a major project to remove exotic creepers and replant native species in order to support the local flora. The clearing took place at the Nature Reserves.

In 1997, a major project lasting about a year was carried out by NParks to remove exotic species that had spread extensively along a 3km stretch of rainforest at the edge of the Nature Reserves. The aggressive creepers were strangling the mature trees and stunting growth of saplings. In some cases, trees were covered entirely with creepers. When lightning struck, trees would fall because they are entangled together by these creepers, hence causing a "domino effect". These creepers consist of exotic species such as *Dioscorea* and *Mikania micrantha*. Once there is a gap, exotic species such as rubber trees and *Clidemia hirta* establish themselves immediately. Thus, there is an urgent need to clear the aggressive creepers before they penetrate into the rainforest. Trees that were badly mutilated by creeper growth were removed, after which students and volunteer groups were taught how to do replanting and started work on reforestation. Only native plants were replanted.

Other areas in the Nature Reserves are now constantly monitored for any invasive species that might adversely affect the native ones. Patches of forest have been adopted by schools to assist NParks in carrying out long-term maintenance, and this is also a good opportunity for the students to learn about ecology and get first-hand experience in management of ecosystems. Students also used this as an opportunity to do simple research on reforestation techniques. A booklet was produced to reflect the findings of the research, and serve as educational material to enhance public awareness on the protection of our remaining natural rainforest.

Edited from a contribution to the "Convention on Biological Diversity" prepared by Singapore

CASE STUDY 5.44   Eradication Programmes against the American Mink in Europe

The American mink (*Mustela vison*) has been imported into Europe since the 1920s for fur farming and for deliberate introduction into the wild. The species' present range covers a large portion of eastern and northern Europe. It represents a major threat for the endangered European mink (*Mustela lutreola*) and it also affects many bird populations, especially on islands.

Baltic Sea

The American mink has colonized almost the entire Finnish and Swedish archipelagos in the last few decades, severely affecting native bird communities. Several control programmes have been planned in order to mitigate the impact of the alien species.

In Sweden, experimental eradications have been achieved in several areas to test for efficiency and to monitor effects on bird reproductive success.

In a group of islands of the Archipelago National Park in south-western Finland, with a total area of 12 x 6 km, a mink eradication project was carried out with the aim of restoring local bird populations. Minks were hunted with a portable leaf-blower (normally used to collect fallen leaves) and trained dogs. After the dog identifies a mink hiding-place, the leaf-blower is used to force the animal out of the burrow. In the first year, 65 minks were taken, and an average of 5-7 animals in the following years. Since 1998, no minks have been trapped and the eradication is considered successful. Many bird populations have increased after the control programme, including black guillemot, velvet scoter, tufted duck, mallard and black-headed gull. No response was recorded in populations of the common eider, greylag goose, common merganser and large gulls. The short distance from the mainland and other islands, as well as the winter freezing of the Baltic Sea, makes re-colonization by mink possible. Therefore, permanent monitoring and control are critical.

In Estonia, a mink eradication project was successfully completed on Hiiumaa Island (1000 km^2) with the aim of re-introducing the European mink on the island. The local population originated from animals escaped from a breeding farm that has now closed. During the campaign, 52 minks were trapped using 10 leg-hold traps, and success of the eradication was monitored through collection of mink presence signs in the breeding season. Re-colonization seems unlikely, since the island is 22 km distant from the mainland. A highly trained staff of 1-3 people carried out the campaign during each season, with the co-operation of local operators. Total cost of the intervention was estimated at 70,000-100,000 Euros. The UK government, the Darwinian Initiative for Biodiversity Foundation, and the Tallinn Zoo provided funds. A similar campaign is now planned on the second largest island of Estonia (Saaremaa, 2,500 km^2) for the same purpose.

Iceland

In Iceland the American mink has been established since 1937 and is now present throughout the country. In recent years, several studies have been conducted to assess the feasibility of a total eradication of the species from the country, but no final decisions have been taken as yet. If such a program is carried out, it will be the largest vertebrate eradication ever carried out in Europe.

Edited from Piero Genovesi (2000): Guidelines for eradication of terrestrial vertebrates: a European contribution to the invasive alien species issue. A report prepared on behalf of the "Convention on the Conservation of European Wildlife and Natural Habitats".

Chapter 6

HOW TO USE THIS TOOLKIT

In practice the application and use of this toolkit is going to depend upon the needs of each reader, the needs of their country, the capability within the country and the relevance of the contents to the country. It is inappropriate to prescribe how it should be used, but some suggestions can be made.

The toolkit was designed and written with a global audience in mind, hence inevitably will fully satisfy no one. Extremes will vary from large relatively wealthy countries with policy, infrastructure and resources in place to tackle invasive species – Australia, New Zealand, South Africa and the USA stand out as more advanced than most others in this regard – to those small island developing states with resources stretched to their limits, so graphically described in Case Study 1.2 "Particular Problems Related to Invasive Species in the South Pacific". In the former group, conservation managers have many information resources available to them and are likely to find the toolkit useful to dip into with regard to particular topics, to refresh their ideas, and gain pointers to explore further through other resources. In the latter group, a conservation manager is likely to find that the toolkit in isolation only gives a glimpse of what is needed and much more local support and adaptation together with appropriate national and/or external partners will be needed to make it effective.

In any future activities of the GISP, we anticipate that pilot projects will be needed to work with individual countries or small groups of neighbouring countries with common invasive species problems, and management challenges, in order to adapt, expand and regionalise the toolkit to strengthen its effectiveness. As it stands, the toolkit is a global attempt to summarise best practices with regard to prevention and management of invasive alien species. Sometimes it will include the key information needed to address particular problems, and other times it may point the user to a reference or website. Each country will have specific priorities and problems that may or may not be adequately addressed by the toolkit. Those countries or areas most challenged by invasive alien species will need to validate the toolkit, testing it against their needs, and in the process develop their own version of the toolkit.

This would be an interactive process, and some of the outputs should be fed back into the global toolkit to increase its value, through additional case studies, expanded information sources, knowledge gleaned from local hands-on experience, and so forth. How might this proceed? Let us take a hypothetical country of small islands; they are rich in biodiversity, relatively affluent due to tourism, well informed regarding many of the issues relating to invasive alien species, but need to develop and implement targeted national and regional action plans to address them. This may sound rather like one of the Mascarene Islands countries, but no direct correlation should be assumed. Development of the toolkit might include steps along these lines:

1.   One or two individuals are identified to co-ordinate and stimulate consideration of the toolkit nationally, and they would lead the following process.

2.   Copies of the global toolkit are made available to relevant conservation managers, NGOs, scientists, etc. Depending upon how extensive the group is, selection of representative individuals may be necessary. These people are tasked to assess how they would use the toolkit, how they would like to use the toolkit, what it lacks, and what would be most important to make it useful and relevant to them.

3.   A small national workshop is convened involving the pilot users and external advisers, familiar with the toolkit and the problems associated with prevention and management of invasive alien species. These may come from one of the GISP partners, or a neighbouring country, or be suggested by the ISSG, or be otherwise located. The workshop will review the toolkit content, identify areas where it needs to be changed for local use, and try to extrapolate from this to other countries with similar problems. Tasks, such as the drafting of new sections, expansion of relevant information by adding local sources, securing copies of key information sources, new local case studies, procuring and interpreting local legislation or other relevant local documents, etc. in order to improve the toolkit and make it more relevant to the country would be identified, and tasks allocated.

4.   These tasks would be completed, using local expertise as far as possible, and external assistance where warranted.

5.   Feedback and outputs from other parallel validation exercises would be circulated and incorporated as appropriate. Outputs would also be made available to the global toolkit, which would incorporate any material that made it more comprehensive, through new Case Studies, Annexes, etc.

6.   Pilot projects would also be identified and prioritised at the national workshop and developed in subsequent follow-up activities, again using local or external expertise as most appropriate. These pilot projects would likely focus on specific aspects identified from the review of the toolkit, as perhaps the most urgent, important, or neglected for that country, where the development of national capacity is critical. Assistance, whether from national or international sources, might be needed to develop the national capacity to prepare project proposals to secure funding for these pilot projects.

7.   The revised toolkit and pilot projects would be presented at a further national or, if appropriate, regional workshop. The pilot projects would be implemented, and national capacity to manage invasive alien species would have been significantly enhanced.

In the worst-case scenario, such as a small island state with staff limitations and over-worked trained professionals, often accompanied by a lack of access to the internet and other resources it may be that the invasive alien species problem will have to be dealt with by one person, sometimes as part of other duties. With these challenges in mind, this toolkit is intended to provide an overview for that manager, helping her or him to recognise what they may not be familiar with, their limitations, what help they need, and provide insight into some of the approaches they could take to address the problem. It may also help to identify regional or international sources of advice and assistance. It should provide guidance on how to develop national awareness and raise the profile of invasive alien species issues on the political agenda so that more realistic resources might be made available.

In preparing this toolkit, it was intended that at least, in part, it would be directly relevant to all conservation managers. It should assist them in the design of their work, and encourage utilization of linkages with information resources. It is intended to assist other concerned and affected people in order to improve the response to invasive alien species problems in their respective countries. We hope that it will assist in preventing the homogenisation and loss of the world's biodiversity.

LOCALITY INDEX

Locality Index

TAXA INDEX

Taxa Index

Taxa Index

KEY PUBLICATIONS OF GISP

Mooney, H.A. and R. J. Hobbs (eds). 2000. *Invasive Species in a Changing World*. Island Press, Washington, D.C.

Perrings, C., M. Williamson, and S. Dalmazzone (eds.). 2000. *The Economics of Biological Invasions*. Cheltenham, UK, Edward Elgar Publishing.

Shine, C., N. Williams, and L. Gundling. 2000. *A Guide to Designing Legal and Institutional Frameworks on Alien Invasive Species*. IUCN Gland, Switzerland, Cambridge and Bonn.

Lowe, S. M., Browne, S. Boudjelas, and M. DePoorter. 2001. *100 of the World's Worst Invasive Alien Species, a selection from the Global Invasive Species Database*. IUCN-ISSG, Auckland, New Zealand.

McNeely, J.A., H.A. Mooney, L.E. Neville, P. Schei, and J.K. Waage (eds.) 2001. *A Global Strategy on Invasive Alien Species*. IUCN Gland, Switzerland, and Cambridge, UK.

McNeely, J.A. (ed.). 2001. *The Great Reshuffling: Human Dimensions of Invasive Alien Species*. IUCN, Gland, Switzerland and Cambridge, UK.

Wittenberg, R. and M.J.W. Cock (eds.) 2001. *Invasive Alien Species: A Toolkit of Best Prevention and Management Practices*. CAB International, Wallingford, Oxon, UK.

Mooney, H.A., J.A. McNeely, L.E. Neville, P.J. Schei and J.K. Waage (eds). *Invasive Alien Species: Searching for Solutions*. Island Press, Washington, D.C. (volume in preparation).

Ruiz, G., and J. T. Carlton (eds). *Pathways of Invasions: Strategies for Management across Space and Time*. Island Press, Washington, D.C. (volume in preparation).